British Poultry Standards

Complete specifications and judging points of all Standardized Breeds and Varieties of Poultry as compiled by the Specialist Breed Societies and recognized by the Poultry Club of Great Britain

Fourth Edition

Edited by **C.G. May**
formerly editor of Poultry World

Revised by **David Hawksworth**
Chairman, Poultry Club of Great Britain

Butterworth Scientific
London Boston Singapore Sydney Toronto Wellington

All rights reserved. No part of this publication may be reproduced or transmitted in any form
or by any means, including photocopying and recording, without the written permission of the
copyright holder, application for which should be addressed to the Publishers. Such written
permission must also be obtained before any part of this publication is stored in a retrieval
system of any nature.

This book is sold subject to the Standard Conditions of Sale of Net Books and may not be
resold in the UK below the net price given by the Publishers in their current price list.

First published in 1954 by Iliffe Books, an imprint of the Butterworth Group
Second edition 1960
Third edition 1971
Reprinted 1976
Reprinted 1977
Reprinted 1978
Reprinted 1980
Fourth edition 1982
Reprinted 1988

© **Poultry Club of Great Britain 1982**

British Library Cataloguing in Publication Data

British poultry standards – 4th ed.
  1. Poultry breeds – Great Britain
  I. May, C.G.   II. Hawksworth, David
  636.5′00941   SF485

ISBN 0-408-70952-9

Typeset by Butterworths Litho Preparation Department
Printed in England by Butler & Tanner Ltd., Frome & London

Contents

Introduction   1

Large fowl and bantams   25

Ancona   25
Andalusian   28
Appenzeller   31
Araucana   33
Rumpless Araucana   36
Asil   38
Australorp   42
Autosexing   45
Barnevelder   47
Belgian Bantams   51
Booted Bantams   59
Brahma   61
Bresse   65
Campine   67
Cochin   70
Crève-Coeur   74
Croad Langshan   76
Dorking   79
Faverolles   83
Frizzle   87
Hamburgh   91
Houdan   96
Indian and Jubilee Indian Game   99
Ixworth   106
Japanese Bantams   107
Jersey Giant   111
Kraienköppe   114

Contents

La Flèche   117

Lakenvelder   119

Leghorn   121

Malay   128

Marans   131

Marsh Daisy   135

Minorca   138

Modern Game   142

Modern Langshan   148

Nankin Bantams   151

New Hampshire Red   152

Norfolk Grey   155

North Holland Blue   157

Old Dutch Bantams   160

Old English Game   163

Old English Game Bantams   171

Old English Pheasant Fowl   176

Orloff   179

Orpington   182

Pekin Bantams   188

Plymouth Rock   192

Poland   199

Redcap   204

Rhode Island Red   206

Rhode Island Red Bantams   210

Rosecomb Bantams   212

Rumpless Game Bantams   215

Scots Dumpy   217

Scots Grey   219

Sebright Bantams   222

Shamo   224

Sicilian Buttercup   226

Silkie   229

Spanish   233

Sultan   235

Sumatra Game   237

Sussex   240

Transylvanian Naked Neck   247

Tuzo Bantams   249

Vorwerk   251

Welsummer   253

Wyandotte   257

Yokohama   272

Turkeys   275

Beltsville White   275

Blue   276

British White   278

Buff   280

Cröllwitzer   281

Mammoth Bronze   283

Norfolk Black   284

Ducks   287

Aylesbury   287

Black East Indian   289

Blue Swedish   291

Campbell   293

Cayuga   296

Crested   298

Decoy   299

Indian Runner   302

Magpie   308

Muscovy   309

Orpington   312

Pekin   313

Rouen   315

Contents

Rouen Clair   318

Saxony   321

Silver Appleyard   323

Silver Appleyard Bantams   324

Welsh Harlequin   325

Other breeds   327

Geese   328

African   328

American Buff   331

Brecon Buff   332

Buff Back   334

Chinese   336

Embden   338

Grey Back   340

Pilgrim   341

Roman   342

Sebastopol   344

Toulouse   346

Other breeds   347

Standard for eggs   349

Glossary   353

(including illustrations of types of comb, wing point, feather marking and leg)

Sitters and non-sitters   369

Defects and deformities   371

(including illustrations of head faults, dished bill, faults in comb, back and breast, tail and leg)

Colour plates

Birds

Plate 1 Australorp, male. Dorking, silver grey female. Indian Game, male. Jubilee Indian Game, female. Orpington, buff male. Old English Pheasant Fowl, female. Ixworth, male
    facing page 4

Plate 2 Redcap, male. Light Sussex, female. Old English Game: black-breasted red male, spangled male, wheaten female. Scots Dumpy, female. Scots Grey, male
    facing page 6

Plate 3 Brahma, light female. Cochin, partridge male. Silkie, white male. Croad Langshan, female. New Hampshire Red, female. Jersey Giant, white female. Plymouth Rock, white male. Rhode Island Red, female. Wyandotte, silver laced female
    facing page 8

Plate 4 Ancona, male. Campine, silver male. Leghorn, buff male. Andalusian, female. Leghorn, white female. Hamburgh, silver spangled female. Minorca, black female. Leghorn, brown male
    facing page 10

Plate 5 Barnevelder, double laced male. Bresse, white female. North Holland Blue, female. Houdan, female. Marans, dark cuckoo male. Faverolles, salmon male. Poland, silver female. Welsummer, female
    facing page 12

Plate 6 (Bantams) Rosecomb, black male. Japanese, black-tailed white male. Barbu d'Uccle, millefleur female. Frizzle, white female. Sebright, silver male. Barbu d'Anvers, blue male. Modern Game: black-red male, pile female
    facing page 14

Feathers

4 Plates
    facing pages 16, 18, 20, 22

Acknowledgements

The Poultry Club of Great Britain wishes to acknowledge the following for supplying illustrations:

A. Rice Poultry Photographs
As in previous editions, the majority of the black-and-white illustrations come from the photographic library of Arthur Rice. This is one of the best collections of winning specimens of poultry and waterfowl in this country. Collectively they reflect in no uncertain manner the contribution made to the Standard-bred poultry movement by Arthur Rice. We should also like to thank his son, S. A. Rice, for his help during the compilation of this edition. Photographs of large fowl, bantams and turkeys from *A. Rice Poultry Photographs* appear on pages 27, 29, 43, 49, 56, 60, 63, 67, 69, 73, 77, 81, 85, 88, 89, 92, 93, 97, 100, 103, 105, 107, 109, 113, 120, 123, 124, 125, 129, 133, 140, 143, 146, 147, 150, 153, 156, 159, 165, 166, 173, 174, 183, 185, 187, 189, 193, 195, 197, 198, 200, 203, 205, 208, 213, 219, 220, 223, 227, 230, 234, 236, 238, 241, 242, 243, 245, 248, 255, 258, 259, 261, 262, 270, 271, 273, 277, 279.

Poultry World – for the photographs on pages 115, 118, 137, 161, 216, 218, 225, 252
Max Butler – for the photographs on page 350
The Appenzeller Society – for the photograph on page 32
The Araucana Club – for the photographs on page 35
The British Belgian Bantam Club – for the photograph on page 52
The Rare Breeds Society – for the photographs on pages 38, 40, 177, 178, 180

The British Waterfowl Association wishes to acknowledge *A. Rice Poultry Photographs* for the photographs on pages 294, 308, 311 and 316 (top) and R. J. Whittick, C. J. S. Marler, T. Bartlett and G. Allen for the other photographs of waterfowl.

Introduction

To the commercial poultryman of today, the pure breeds of poultry are of little or no consequence. His outlook penetrates no farther than the hybrid strains of layers and broilers and he either does not know, or has found it convenient to forget, that all of them owe their origin to the standard pure breeds.

Far-seeing geneticists, on the other hand, think differently. Many of them visualize the time when hybrid strains may have to be remade and it will be then that there will come a resurgence in the demand for pure breeds, not necessarily those that are of outstanding show quality but certainly those whose blood lines are pure. Even today, with new hybrids coming off the pipeline, there is a consistent call for pure blood lines.

Standards owe their origin to the demand for uniformity in type and colouration of the various breeds. Without them the shows could never have been born and it was as long ago as in 1865 that the Poultry Club authorized the publication of the first 'Standard of Excellence in Exhibition Poultry', an exceedingly modest predecessor to the larger work which was not to appear until after the turn of the century.

Right from those early days, the Club has remained the guardian of the standards without necessarily being the body responsible for framing them. This task is normally undertaken by the specialist breed club or by the originator of a new breed or variety. So seriously, however, is this guardianship imposed, and accepted by the clubs, that until a new variety is admitted to standard it remains unrecognized by show authorities whose events are staged under the rules of the Poultry Club of Great Britain.

As a result of the incorporation of the British Bantam Association, the Poultry Club finds itself with added responsibilities, particularly by reason of the fact that at most of the shows bantams now outnumber large fowl. There has also been a considerable increase in the number of specialized bantam shows.

Normal procedure for the admittance of a new breed or variety of an existing breed to standard involves the submission to the Poultry Club by the originator of that breed of live specimens of more than one generation of the birds to be standardized, a draft of the proposed standard and a sworn declaration to the effect that the breed reproduces its like to a remarkably high degree. Through its Council the Club considers the breed

Introduction

and recommends changes in the submitted standard where changes are necessary or, in some cases, a complete revision. So searching are these investigations that seldom does it occur that a breed is accepted on first application.

The only deviation from this programme is made when a recognized breed is imported from another country in which it has already been accepted to standard – the Club usually accepts these without the necessity of submitting specimens as sworn declarations.

This is the fourth edition of British Poultry Standards. In it are a number of major revisions/additions without excluding those features which claimed wide popularity for the earlier works. The format has been simplified and both large fowl and bantam standards appear together for each breed.

Retained are the coloured sections featuring no fewer than 47 breeds of large fowl and bantams in their natural colours and four plates of feathers copied from specimens taken from prominent show winners. While each of the feathers is identified to its breed (or breeds), the actual colour descriptions of them in the appropriate breed standard can also be applied to colour descriptions of other breeds. Mahogany, for instance, is the same in Indian Game as in gold laced Wyandottes or millefleur Belgian bantams; the salmon of the brown Leghorn is also the salmon of the Faverolles, and no difference exists between the beetle green of the Ancona and that of the black Leghorn.

As a consequence, therefore, the feathers serve generally as a complete colour guide to every one of the breed standards included in this publication. Accompanying each perfect standard feather is an imperfect one of a type which most often occurs in the breeds specified.

Supplementing the full standards of all the recognized breeds, previous editions of this work contained, in addition, short descriptions of breeds almost on the verge of extinction but still occasionally seen on the show bench. The same applies to this edition with the important exception that, where a breed has revealed unmistakable signs of revival, it has been taken out of the 'Other Breeds' section and given more prominence with a full standard.

Conversely, and where popularity has waned, other breeds have been relegated to the 'Other Breeds' section. Where a standardized breed, or variety of a breed, is now known no longer to exist that breed or variety has been omitted. The work of the Rare Breeds Society has been invaluable in reviving interest in, and finding standards for, many breeds which otherwise would be extinct. The Society holds the standards for most of the rarer varieties sometimes seen in this country but not included in this book.

Introduction by the Club of a judges' panel has gone a long way towards standardizing judging. Under a system of both practical and written examination judges are graded according to their capabilities. Newcomers are probably satisfied to qualify for a single breed only, coming up for further examination as their experience widens. The 'plum' certificate is, of course, one which qualifies its holder as an all-round judge. Very few of these are awarded.

While every care has been taken in compiling the standards the form of presentation does not necessarily follow that adopted by the specialist

Introduction

breed clubs. Instead, and for ease of reference, a prescribed pattern has been set without in any way departing from the salient points of any of the breeds concerned. It is this form of layout that has the approval of the Poultry Club and is, therefore, recommended to all breed clubs to follow.

To safeguard publication interests the Poultry Club has agreed not to accept or authorize publication of any alterations to existing standards for a period of two years from the issue of this edition. The Poultry Club, through its 53 affiliated breed clubs, maintains the strictest watch on these standards of excellence. It will not allow alterations or amendments until its governing council has made a thorough examination of all the circumstances. In this way the Poultry Club can be truly said to be the guardian of the standards and so play its part in ensuring that our pure breeds of poultry will be part of the heritage we pass on to future generations.

Plate 1

Large fowl

1   Australorp, male
2   Dorking, silver grey female
3   Indian Game, male
4   Jubilee Indian Game, female
5   Orpington, buff male
6   Old English Pheasant Fowl, gold female
7   Ixworth, male

Plate 1

Plate 2

Large fowl

1 Redcap, male
2 Sussex, light female
3 Old English Game, black-breasted red male
4 Old English Game, spangled male
5 Old English Game, wheaten female
6 Scots Dumpy, female
7 Scots Grey, male

Plate 2

Plate 3

Large fowl

1 Brahma, light female
2 Cochin, partridge male
3 Silkie, white male
4 Croad Langshan, female
5 New Hampshire Red, female
6 Jersey Giant, white female
7 Plymouth Rock, white male
8 Rhode Island Red, female
9 Wyandotte, silver laced female

Plate 3

Plate 4

Large fowl

1 Ancona, male
2 Campine, silver male
3 Leghorn, buff male
4 Andalusian, female
5 Leghorn, white female
6 Hamburgh, silver spangled female
7 Minorca, black female
8 Leghorn, brown male

Plate 4

Plate 5

Large fowl

1 Barnevelder, double laced male
2 Bresse, white female
3 North Holland Blue, female
4 Houdan, female
5 Marans, dark cuckoo male
6 Faverolles, salmon male
7 Poland, silver female
8 Welsummer, female

Plate 5

Plate 6

Bantams

1 Rosecomb, black male
2 Japanese, black-tailed white male
3 Barbu d'Uccle, millefleur female
4 Frizzle, white female
5 Sebright, silver male
6 Barbu d'Anvers, blue male
7 Modern Game, black-red male
8 Modern Game, pile female

Plate 6

Plate 7

Standard feather markings

1 Hackle feather conforming to standard as applying to brown Leghorn and other males of black-red colouring. Note the absence of shaftiness, black fringing and tipping. Actual colour of outer border varies in different breeds between dark orange and pale lemon. In such breeds saddle hackle should conform closely to neck hackle.

1A Faulty hackle in same breeds. There is considerable shaftiness, the striping runs through and the feather is tipped with black. Striping is also indefinite and fouled with red.

2 Hackle feather conforming to standard from partridge Wyandotte male. There is no shaftiness and the striping is very solid and distinct. In partridge Wyandottes lemon-coloured hackles are a desirable exhibition point.

2A Faulty neck hackle in the same breed. Note that the black striping runs through to tip and is irregular in shape. There is also a distinct black outer fringing to the gold border.

3 Standard hackle feather from male of gold-laced Wyandotte and similar breeds with rich bay ground colour. Note intensity of centre stripe, absence of shaftiness and freedom from blemish in outer border. Note also soundness of colour in underfluff.

3A Faulty hackle feather from similar breeds, showing indistinct striping, with foul colour, shaftiness and black running through to tip. Underfluff is a mixture of red and dark grey.

4 Standard hackle feather from male of light Sussex and similar breeds of ermine markings, such as light Brahma, columbian Wyandotte and ermine Faverolles. The demand is for solid black centre with clear white border extending to underfluff. Green sheen is an important feature.

4A Faulty hackle feather from similar breeds, showing black fringing to border, black tipping and shaftiness in quill. Underfluff also lacks distinction.

5 Perfect tri-coloured hackle feather from a speckled Sussex male. The black striping is solid, with green sheen, and the border is the desired rich mahogany colour, finishing with clean white tip. Note clarity of undercolour.

5A Faulty speckled Sussex hackle feather showing almost complete lack of black striping, varying ground colour in border, and indistinct white tipping.

6 Neck hackle conforming to standard of Andalusian male. The so-called Andalusian blue is a diffusion of black and white, and in male hackles a dark border or lacing surrounds the slate blue feather. Undercolour is sound and even.

6A Faulty hackle from same breed. The colour generally is blotchy and lacing is indefinite.

7 Standard neck hackle of a Rhode Island Red male. No attempt has been made to show the ultra-dark red usually seen in show specimens, but the colour seen here conforms with standard and should be agreeable for exhibition. Note purity of undercolour – a very important point in this breed.

7A Faulty hackle feather from the same breed, showing uneven ground colour, black tipping and smutty undercolour, which is a very severe defect in a Rhode Island Red.

8 Hackle from Ancona male, conforming closely to standard. Note clear V-shaped white tipping, complete absence of shaftiness, rich green sheen and solidity of dark underfluff, a particularly strong point in the breed.

8A Faulty hackle feather from same breed, showing indistinct tipping of greyish white and faulty undercolour not dark to skin.

9 Hackle feather conforming to standard from buff Orpington male, very similar, except for exact shade, to feathers from other buff breeds, such as Cochins and Rocks. Note even colour throughout, absence of shaftiness and sound colour in underfluff, with quill buff to skin.

9A Faulty hackle feather from similar breed, showing severe shaftiness, uneven ground colour with darker fringe, and impure undercolour.

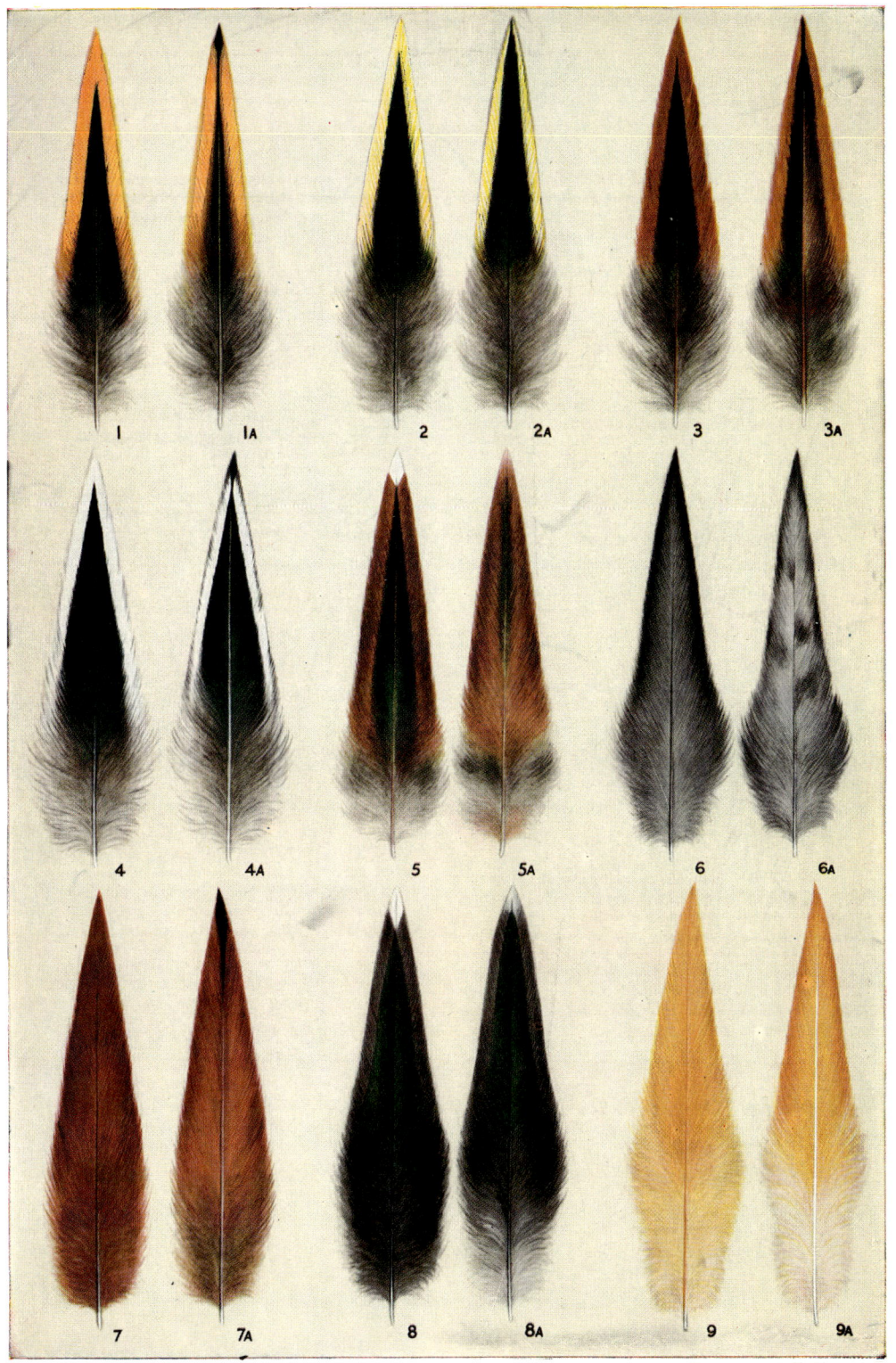

Plate 7

Plate 8

Standard feather markings

1 Standard hackle from barred Plymouth Rock male and similar breeds. Note the points of excellence – barring practically straight across feather, sound contrast in black and blue-white, barring and ground colour in equal widths, and barring carried down underfluff to skin. Tip of feather must be black.

1A Faulty saddle or neck hackle from similar variety. There is lack of contrast in barring, with dull grey ground colour and V-shaped bars.

2 Hackle as standard description from silver Campine, in which males are inclined to hen feathering. Note that the black bar is three times the width of ground colour and tip of feather is silver.

2A In this faulty hackle (also from silver Campine male) ground colour is too wide and barring narrow. Feather is without silver tip.

3 Standard hackle from Marans male. In this and some similar breeds evenness of barring is not essential, but it is expected to show reasonable contrast. It should, however, carry through to underfluff.

3A From the same group of breeds this feather is far too open in barring and lacks uniformity of marking. It is also light in undercolour.

4 Standard markings of female body feather in Plymouth Rocks and similar barred breeds where barring and ground colour are required to be of equal width. Note that barring runs from end to end of feather and that tip is black.

4A Faulty feather from same group. Note absence of barring to underfluff and V-shaped markings; also blurred and indistinct ground colour.

5 Sound body feather from silver Campine female showing standard silver tip and barring three times as wide as ground colour, as in the male. Gold Campine feathers are similar but for difference in ground colour.

5A Faulty female feather, again from silver Campine. Here again, as in 2A, barring is too narrow in relation to silver ground colour and tip of feather is black.

6 Body feather from Marans female, conforming to standard requirements. Note that the markings are less definite than in Rocks and Campines, and the black is lacking in sheen, while ground colour is smoky white.

6A Faulty Marans female feather. Lacks definition and contrast in barring, which is indefinite in shape, the blotchy ground colour making an indistinct pattern.

7 Excellent body feather from partridge Wyandotte female, showing correct ground colour and fine concentric markings. Note complete absence of fringing, shaftiness and similar faults. Fineness of pencilling is a standard requirement.

7A From the same breed this faulty female feather shows rusty red ground colour and indistinct pencilling, with faulty underfluff.

8 Body feather of standard quality from Indian or Cornish Game female. The illustration shows clearly two distinct lacings with a third inner marking. Lacing should have green sheen on a rich bay or mahogany ground.

8A Faulty feather from same breed. Missing are evenness of lacing and central marking. The outer lacing runs off into a spangle tip.

9 Standard feather from laced Barnevelder female. In this breed ground colour should be rich with two even and distinct concentric lacings. Quill of feather should be mahogany colour to skin.

9A Faulty Barnevelder female feather, showing spangle tip to outer lacing and irregular inner markings on ground colour that is too pale.

Plate 8

Plate 9

Standard feather markings

1 Standard markings on silver laced Wyandotte female feather, showing very even lacing on clear silver ground colour and rich colour in underfluff. In this breed clarity of lacing is of greater importance than fineness of width.

1A Faulty female feather from same breed. In this there is a fringing of silver outside the black lacing, which is irregular in width and runs narrow at sides. Undercolour is also defective.

2 Excellent feather from gold laced Wyandotte. In this the ground colour is a clear rich golden bay and there is a complete absence of pale shaft. Undercolour is sound and lacing just about the widest advisable.

2A This shows a very faulty feather from same breed. It portrays mossy ground colour with blotchy markings and uneven width of lacing at sides of feather. Undercolour is not rich enough.

3 Standard markings on Andalusian female feather showing well-defined lacing on clear slate blue ground and good depth of colour in underfluff. The dark shaft is desirable and is not classed as a fault.

3A Faulty feather from female of same breed. In this the ground colour is blurred and indistinct, and the lacing is not crisp, while undercolour lacks depth.

4 This shows a feather from an Ancona female, almost perfect in standard requirements. The white tipping is clear and V-shaped and undercolour is dark to skin.

4A Faulty feather from female of same breed. Here the tip of feather is greyish white and lacks the necessary V-shape, while undercolour is not rich enough.

5 An almost perfectly marked feather from a speckled Sussex female – though the white tip might be criticized by some breeders as rather too large. The black dividing bar shows good green sheen and the ground colour is rich and even.

5A As a contrast this faulty feather shows a blotchy white tip and lack of colour in underfluff. The ground colour is also uneven.

6 An excellent example of 'mooning' on the feather of a silver spangled Hamburgh female. Note the round spangle and the clear silver ground with sound undercolour.

6A In this feather from the same breed the spangling at tip is not moon-shaped and there is too much underfluff and insufficient silver ground colour to body of feather.

7 A good example of the desired colour in Rhode Island Red female plumage. Note the great depth of rich colour and the sound dark undercolour.

7A Faulty colour in a feather from the same breed. Here the middle of feather is paler and inclined to shaftiness, and colour generally is uneven.

8 Standard plumage in females of Australorp and similar breeds of soft feather with rich green sheen. Note the brilliance of colour and general soundness of underfluff.

8A This shows a common fault in similar breeds, a sooty or dead black colour without sheen and lacking lustre. This sootiness is, however, usually accompanied by dark undercolour.

9 Standard colour and feather in the buff Rock female and similar breeds which perhaps vary in exact shade and in quantity and softness of underfluff. Note the clear even buff and lack of shaftiness or lacing, also the sound rich undercolour.

9A This feather from a similar buff breed shows very bad faults – mealiness and bad undercolour with a certain amount of pale colour in shaft.

Plate 9

Plate 10

Standard feather markings

1 This shows a typical standard bred feather from a Derbyshire Redcap female. Note the rich ground colour and the crescentic black markings, which are really midway between spangling and lacing.

1A In this faulty feather from a female of the same breed the ground colour is uneven and lacks richness, while the black tip is too small and indefinite and too closely resembles moon-shaped spangling.

2 This is a standard example of the webless type of plumage associated with Silkies in which the feather vane has no strength and the barbs no cohesion. This plumage is common in all colours.

2A Faulty feather from the same breed. In this the middle of feather is too solid and lacks silkiness, while the fluff has insufficient length.

3 A delicately pencilled body feather from a silver grey Dorking female. Note the silvery colour and absence of ruddy or yellow tinge in ground colour. This type of feather is also usual in duckwing females of various breeds.

3A Faulty colour in female feather from same breed. Here there is a distinctly incorrect ground colour and pronounced shaftiness.

4 A good example of standard bred colour and markings in body feather of brown Leghorn female, where the ground colour is a soft brown shade and the markings finely pencilled. This type of feather is common to many varieties of partridge or grouse colouring.

4A This shows a body feather from the same breed, in which ground colour is ruddy and shaftiness is pronounced – both severe exhibition faults.

5 A well-chosen example of the irregularity in markings of an exchequer Leghorn female. In this breed the black and white should be well distributed but not regularly placed, and underfluff should be parti-coloured black and white.

5A This faulty feather from the same breed shows a too regular disposition of markings, the body of the feather being almost entirely black and the white markings almost resembling lacing.

6 This is a standard feather from the breast of a silver Dorking, and with slight variations of shade from pale to rich salmon applies to a number of varieties with black-red or duckwing colouring. Colour should be even with as little pale shaft as possible.

6A A faulty sample of breast feather from the same group. Here the ground colour is washy and disfigured by pale markings known as mealiness.

7 Standard markings in North Holland Blue female. Note the defined but somewhat irregular barring on a distinctly bluish ground. No barring or other requirements in underfluff are called for in the standard.

7A This shows a faulty female feather in the same breed, which is not closely standardized for markings. The ground colour is smoke grey instead of blue, and is blotchy, with uneven markings.

8 A good example of clear colour in an unlaced or self-blue female feather, where no lacing is permissible, such as in blue Leghorns, blue Wyandottes, etc. Note even pale blue shade and absence of any form of markings. This is an example of the true-breeding blue colour found in Belgian bantams.

8A This faulty female feather is a dull dirty grey instead of clear blue, and has blotchy markings as well as a suggestion of irregular lacing.

9 A good sample of exquisitely patterned thigh fluff in Rouen drakes. The ground colour is a clear silver and the markings a delicate but clear black or dark brown. These markings are sometimes known as chain mail.

9A Another good Rouen feather – this time from the duck. Ground colour is very rich and markings intensely black, though seldom so regular and even as in domestic fowl.

Plate 10

British Poultry Standards

This composite drawing is designed to illustrate the chief points of the various breeds of fowl

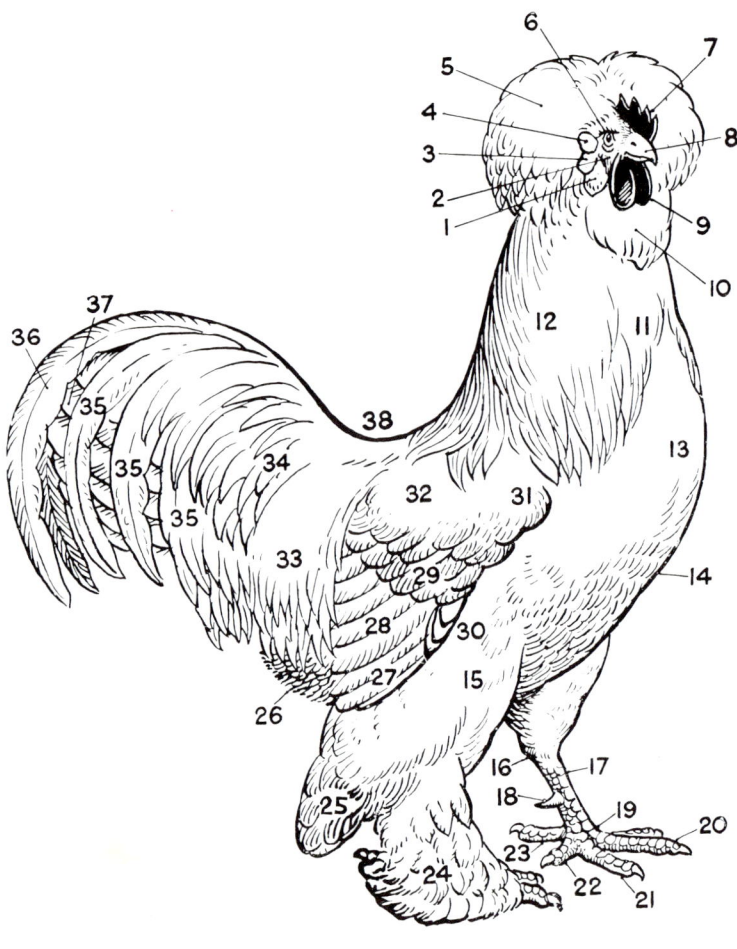

1	Muffling	14	Keel	27	Primary flights
2	Face	15	Thigh	28	Wing bay
3	Ear-lobe	16	Hock joint	29	Wing bar
4	Ear	17	Shank	30	Wing covert
5	Crest	18	Spur	31	Shoulder
6	Eye	19	Foot	32	Wing bow
7	Comb	20	Middle toe	33	Saddle hackle
8	Beak	21	Third toe	34	Tail coverts
9	Wattles	22	Fourth toe	35	Side hangers
10	Beard	23	Fifth toe	36	Tail sickle
11	Neck	24	Footings	37	Main tail
12	Neck hackle	25	Vulture hock	38	Back
13	Breast	26	Abdomen		

Large fowl and bantams

Ancona

Large fowl

Origin: Mediterranean
Classification: Light
Egg colour: White to cream

Named after the province of Ancona in Italy, specimens of this Mediterranean breed were imported into England in 1851, first the single then the rose comb. Controversy centres around the view that Anconas are akin to the original mottled Leghorn and, therefore, a member of the Leghorn family. However, the fact remains that breeders adhere to the name of Ancona. The breed has retained its popularity on the show-bench not only for its laying propensities, but because of its combination of breed type and characteristics with usefulness.

General characteristics: male

Carriage: Alert, bold and active.

Type: Body broad, close and compact. Back of moderate length. Breast full and broad, carried well forward and upward. Wings large and carried well tucked up. Tail full and carried well out.

Head: Deep, moderate in length, rather inclined to width, and carried well back. Beak medium with a moderate curve. Eyes bright and prominent. Comb single or rose. The single of medium size, upright and with five to seven deep broad and even serrations forming a regular curve, coming well back and following the line of the head, free from excrescences. The rose resembles that of the Wyandotte. Face smooth and of fine texture. Ear-lobes medium, inclined to almond shape, free from folds. Wattles long, fine in texture, in proportion to comb.

Neck: Long, nicely arched and well covered with hackle.

Legs and feet: Legs of medium length, strong, set well apart, clear of feathers, thighs not much seen. Toes, four, rather long and thin and well spread out.

Ancona

Female

With the exception of the single comb, which falls gracefully to one side of the face, without obscuring the vision, the characteristics are generally similar to those of the male, allowing for the natural sexual differences. The body, however, is round and compact, with greater posterior development than in the male. The back is rather long and broad, the neck of medium length and carried well up.

Colour

Male and female plumage: Good beetle green ground tipped with white (the more V-shaped the better). No inclination to lacing. The more evenly V-tipped throughout with beetle green and white the better, provided the ground colour is beetle green.

In both sexes: Beak yellow with black or horn shadings; a wholly yellow beak not desirable. Eyes, iris orange-red, pupil hazel. Comb, face and wattles bright red, face free from white. Ear-lobes white. Legs yellow, mottled with black, the more evenly mottled the better.

Weights

Cock 2.70–2.95 kg (6–6½ lb); cockerel 2.50 kg (5½ lb)
Hen 2.25–2.50 kg (5–5½ lb); pullet 2.00 kg (4½ lb)

Scale of points

Type and carriage	15
Texture (general)	10
Size	5
Purity of white, quality and evenness of tipping	20
Beetle green ground colour dark to skin	15
Leg colour	5
Head (eye 5, comb 10, lobe 5)	20
Beak colour	5
Condition	5
	100

(*Note:* Eye points include brightness and prominence. Comb points include medium size and fine texture.)

Defects

	To lose
In-kneed	10
Squirrel tail	10
Crooked toes	10

(*continued*)

Ancona

**Ancona large fowl
Singlecomb male
Rosecomb female**

Andalusian

(**Defects** – *continued*)

White or light undercolour	10
Ground colour other than beetle green	10
Tail not tipped or not black to roots	10
Wings any other colour than black tipped with white	10
Bad comb	5
White in face	20
Lobe other than white	5
	100

(*Note:* Roach back or any bad structural deformity a disqualification.)

Bantams

Ancona bantams are to be exact miniatures of their large fowl counterparts and so standard, colour and scale of points are to be used as for large fowl.

Weights

Male   570–680 g (20–24 oz)
Female 510–620 g (18–22 oz)

Originally British Bantam Association weights were fractionally higher for males.

Andalusian

Large fowl

Origin: Mediterranean
Classification: Light
Egg colour: White

The breed owes its name to the Province of Andalusia in Spain, and is one of the oldest of the Mediterranean breeds. It is a contemporary of the black Spanish with which, no doubt, it is closely related. The blue Andalusian, as we know it today, was developed from black and white stock imported from Andalusia about 1846, and blending of the two colours most probably created the blue. The earlier specimens were large, and game-like in carriage, with medium combs and lobes, and of a self colour, although individual birds were selectively bred for lacing, by infusion of black Minorca blood.

Andalusian

Andalusian large fowl
Blue male
Blue female

Andalusian

General characteristics: male

Carriage: Upright, bold and active.

Type: Body long, broad at the shoulders, and tapering to the tail, with the plumage close and compact. Breast full and round. Wings long, well tucked up and the ends covered by the saddle hackles. Tail large and flowing, carried moderately high, but not approaching 'squirrel' or fan shape.

Head: Moderately long, deep and inclined to width. Beak stout and of medium length. Eyes prominent. Comb single, upright and of medium size, deeply serrated with spikes broad at the base, the back portion slightly following the line of the head but not touching the neck. Free from 'thumb marks' or side spikes. Face smooth. Ear-lobes almond in shape, medium size, free from wrinkles, and fitting closely to the face. Wattles fine and long.

Neck: Long and well covered with hackle feathers.

Legs and feet: Legs long. Shanks and feet free from feathers. Toes, four, straight and well spread.

Female

With the exception of the comb, which falls with a single fold to one side without covering the eye, the general characteristics are similar to those of the male, allowing for the natural sexual differences.

Colour

Male and female plumage: Clear blue, edged with distinct black lacing, not too narrow, on each feather, excepting the male's sickles, which are dark (or even black), and his hackles, which are black with a rich gloss, while the female's neck hackle is a rich lustrous black, showing broad lacing on the tips of the feathers at the base of the neck. Undercolour to tone with surface colour.

In both sexes: Beak dark slate or horn. Eyes dark red or red-brown. Comb, face and wattles bright red. Ear-lobes white. Legs and feet dark slate or black.

Weights

Male   3.20–3.60 kg (7–8 lb)
Female 2.25–2.70 kg (5–6 lb)

Scale of points

Ground colour	30
Lacing	20
Head (comb 10, face 10, lobes 5)	25
Size, type, carriage, tail and condition	25
	100

Serious defects

In the male, much white in face or presence of red in lobes. White feathers. Sooty ground colour. Red or yellow in hackles. Any deformity and comb not upright. In the female any of these points that apply, together with an upright comb.

Bantams

Andalusian bantams are exact miniatures of their large fowl counterparts and so standard, colour and scale of points apply.

Weights

Male   680–790 g (24–28 oz)
Female 570–680 g (20–24 oz)

Appenzeller

Origin: Switzerland
Classification: Light
Egg colour: White

The Appenzeller has recently been imported into this country from the Continent. It has been popular and has attracted enough supporters to warrant forming a specialist club to look after the interest of the breed. There is no Appenzeller bantam standardized at the moment.

General characteristics: male

Carriage: Neat, and very active.

Type: Body well rounded, medium long, walnut-shaped. Breast full, carried high. Wings rather long, carried close. Tail well furnished, well spread, at right angles to back. Abdomen well developed. Back of medium length, slightly sloping with full hackle.

Head: Medium sized, held high, with typical raised skull and medium-sized pointed crest bent forward. Face smooth. Comb horn type consisting of two small rounded spikes, separate and without side sprigs. Wattles moderately long and fine. Ear-lobes medium sized, oval. Beak powerful, with strong cavernous nostrils, and a prominent horseshoe ridge to the beak with a small fleshy knob at the front. Eyes prominent and alert.

Neck: Medium length, slightly arched with abundant hackle.

Legs and feet: Thighs slender and prominent; shanks medium length, fine. Toes, four, well spread.

Plumage: Fairly hard and tight.

Appenzeller

Appenzeller large fowl
Spitzhauben female

Female

Except for a more horizontal back line the general characteristics are the same as for the male, allowing for the natural sexual differences.

Colour

The silver spangled
Male plumage: Pure silvery white ground colour, with each feather ending in distinct black fairly small spangle, not circular; less pronounced on the head and neck. Primaries, secondaries and tail feathers with black tips. Abdomen and fluff grey; underplumage dark grey.
Female plumage: Head crest and the neck silver white with black tipping. Breast, wing bows, back and tail silvery white with distinct black spangling. Flights as for the male. Undercolour dark grey.

The gold spangled
Male plumage: Gold-red ground colour, spangling as for the silver spangled. Flights: outer web golden yellow, inner web as black as possible. Breast and flanks gold with black spangles; abdomen and underplumage greyish black. Tail as brown as possible with black tips, a blackish brown tail allowed.
Female plumage: Golden yellow ground colour, tail golden brown with black spangling; otherwise as for the male, having regard to the necessary sexual differences.

The black
Male and female plumage: Shiny greenish black with dark grey to black underplumage.

In both sexes and all colours: Beak bluish. Eyes dark brown. Comb, face and wattles bright red; ear-lobes bluish white. Shanks blue.

Weights
Male   1.60–2.00 kg (3½–4½ lb)
Female 1.35–1.60 kg (3–3½ lb)

Scale of points
Type and carriage	25
Colour and markings	25
Head points	25
Legs and feet	15
Condition	10
	100

Serious defects
Comb other than horn. Side sprigs. Narrow or roach back. Squirrel tail. Breast too deep or narrow. Low wing carriage. Tail lacking fullness. Crow beak. Nostril not cavernous. Bad stance. Any sign of feathering on shanks.

Araucana

Large fowl

Origin: Chile
Classification: Light
Egg colour: Blue or green

When the Spaniards arrived in South America, bringing with them the light Mediterranean breeds, they found that the indigenous Indians had domestic fowls which soon crossbred with the incomers. Notable for their fierce resistance to the Spaniards however, were the Indians of the Arauca province of northern Chile who were never conquered. The name Araucana for the breed is derived therefore from that part of the world where the South American and European fowls had the least opportunity to interbreed.

The Araucana breed standard in the British Isles is generally as envisaged by George Malcolm who created the true-breeding lavender Araucana, amongst other colours, in Scotland during the 1930s. Araucanas

Araucana

are prolific layers of strong-shelled eggs, blue or green eggs having been reported from South America from the mid sixteenth century onwards. These are unique in that their colour permeates throughout the shell.

General characteristics: male

Carriage: Alert and active.

Type: Body long and deep, free from heaviness. Firm in handling. Back moderately long, horizontal. Wings large and strong. Tail well developed with full sickles carried at an angle of 45°.

Head: Moderately small. Beak strong and stout. Eyes bold. Comb small pea. Face covered with thick muffling and ear muffs abundant. Crest compact, carried well back from eyes. Ear-lobes moderately small and concealed by muffling. Wattles round and very small, hardly perceptible, smooth to touch.

Neck: Of medium length abundantly furnished with hackle feathers.

Legs and feet: Medium length strong and well apart. Toes, four, straight and well spread.

Female

The general characteristics are similar to those of the male allowing for the natural sexual differences. Comb minimal.

Colour

Male and female plumage: Any Old English Game colour excluding off-colours. Any self colour including lavender, both sexes to be a uniform blue-grey throughout.

In both sexes and all colours: Comb, face and wattles bright red. Eyes dark orange in all colours. Legs willow to olive or slate.

Weights

Male   2.70–3.20 kg (6–7 lb)
Female 2.25–2.70 kg (5–6 lb)

Scale of points

Type and carriage	20
Crest and muffling	25
Comb	10
Other head points	5
Feet and legs	5
Colour	20
Condition and handling	15
	100

Araucana

Araucana bantams
Lavender male
Lavender female

Serious defects or disqualifications

Cut-away breast. Roach back. Wry or squirrel tail. Crest too small or too large, e.g. Poland type. Absence of crest or muffling. Comb other than of pea type in male. Comb lopped or twisted. Any comb other than minimal in female. Pearl eye. Legs other than standard colour. Uneven or splashed breast colour. In males white in base of tail. In lavenders any straw or brassy tinge.

Bantams

The standard to be an exact miniature of the large fowl. The quality of bantam Araucanas at present being exhibited is generally higher than that of the large fowl. The most popular colour is lavender. Araucana bantams are extremely productive layers of blue eggs.

Weights

Male   740–850 g (26–30 oz)
Female 680–790 g (24–28 oz)

Serious defects

As large fowl, plus low wing carriage. High tail carriage. Any tendency to rose comb. Scale of points as in large fowl.

Rumpless Araucana

Large fowl

Origin: Chile
Classification: Light
Egg colour: Blue or green

The Rumpless Araucana also has its origins in South America. It was introduced to Europe by Professor S. Castello in the early 1920s. The ear-tufts of feathers are unique to the breed in that they grow from a fleshy pad adjacent to the ear-lobe. The breed standard recently ratified in this country is similar to that currently in use in continental Europe and the U.S.A. Rumpless Araucanas lay a large egg in relation to body size and are as productive as the tailed Araucanas.

General characteristics: male

Carriage: Alert, active and assured.

Type: Body moderate in length, broad at shoulders. Back flat and slightly sloped. Rump well rounded with saddle feathers flowing over stern. Breast full, round and deep. Wings medium in length, carried close to the body and well up. Saddle hackle well developed. Tail entirely absent, with no uropygium.

Head: Moderately small. Beak medium stout, curved. Eyes bold and expressive. Comb small pea. Face moderate muffling. Ear-lobes small and concealed by ear-tufts. These originate from a gristly appendage arising from behind and just below the ear hole. The tufts of feathers, numbering from 5 to 15, grow from this pad. The tufts should be of a good length, matching in size and extending from the ears backwards in a well-defined sweep, or project horizontally. Wattles very small.

Neck: Medium length, well furnished with hackle feathers.

Legs and feet: Medium in length, straight and well set apart. Toes, four, strong and well spread.

Female

The general characteristics are similar to those of the male allowing for the natural sexual differences.

Colour

Male and female plumage: As for Araucanas.

In both sexes and all colours: Eyes dark orange. Legs and feet olive or slate.

Weights

Excess weight to be penalized.
Male   2.70 kg (6 lb)
Female 2.25 kg (5 lb)

Scale of points

Type and carriage	20
Ear-tufts	25
Comb	5
Other head points	5
Feet and legs	5
Colour	15
Condition and handling	25
	100

Serious defects

Non-standard comb. Unmatched ear-tufts. Shape other than standard, e.g. narrow body. Any tail feathers (incomplete rumpless). Fluff showing below saddle hackle.

Disqualifications

No ear-tufts, single ear-tuft, crest.

Asil

Rumpless Araucana
large fowl
White male

Bantams

These should be a true miniature of the large Rumpless Araucana. As the large Rumpless fowl is historically and naturally a small breed, it follows that great care must be taken to keep the bantams within the approved weight limits. Colours at present include black-red, black and white.

Weights

Excess weight to be penalized.
Male   910 g (32 oz)
Female 790 g (28 oz)

Asil

Large fowl

Origin: Asia
Classification: Light
Egg colour: Tinted

The Asil is probably the oldest known breed of game fowl, having been bred in India for its fighting qualities for over 2000 years. The name Asil is derived from the Arabic and means 'of long pedigree'. In its native land the

Asil

Asil was bred to fight, not with spurs, but rather with its natural spurs covered with tape and the fight became a trial of strength and endurance. Such was the fitness, durability and gameness of the contestants that individual battles could last for days. This style of fighting produced a powerful and muscular bird with a strong beak, thick, muscular neck and powerful legs and thighs together with a pugnacious temperament and a stubborn refusal to accept defeat.

Never very numerous in Britain, the Asil has nevertheless always attracted a few dedicated admirers prepared to cope with its inborn desire to fight, a characteristic shared by the females who are poor layers but extremely good mothers.

General characteristics: male

Carriage: Upright, standing firmly and well on his legs. Sprightly and quick in movement. When seen in profile the eye should be directly above the middle toe-nail.

Type: Chest wide and well thrown out. Hard and muscular and feeling remarkably flat in the hand. Back broad and flat tapering to a fairly narrow stern but very well developed and strong at the root of the tail. Viewed from above the body should appear to be heart-shaped. Wings carried well out from the body at the shoulders and should be muscular where they join the body but otherwise carrying very little flesh and covered with hard feathers and tough, rather short quills. Sickle feathers narrow and scimitar-shaped, drooping from the base. Saddle feathers pointing backwards more so than in other breeds.

Head: Skull broad with large and square jaw bones and large cheek bones covered with tough leathery skin. Beak short, thick, powerful, shutting tight. Eyes bright and bold set in oval pointed eyelids. Iris pearl colour but occasionally seen slightly bloodshot or with a yellowish tinge. Comb triple or pea, very hard fleshed and set low. No wattles.

Neck: Medium length carried slightly curved to give a short appearance. Thick and very hard to the touch and covered with short, hard and wiry feathers. Throat clean cut with bare skin extending well down the neck.

Legs and feet: Thick and square with a noticeable indentation down the front of the leg where the scales meet. Toes, four, straight, strong and tapering with broad, curved toe-nails. Thighs not too long, round, hard muscular and when viewed from the front should be in line with the body and not the shoulders.

Plumage: Short and wiry. Difficult to break and with little or no underfluff. Patches of bare skin showing red are to be seen on the breast bone, wing joints and thighs.

Handling: Firm and muscular. Heavier in the hand than appearance would at first suggest.

Asil

Asil large fowl
Male

Asil large fowl
Female

Female

The general characteristics are similar to those of the male allowing for the natural sexual differences.

Colour

Male and female plumage: There are no fixed colours, the principal colours seen today being light red and dark red with grouse-coloured and red wheaten females. Greys, spangles, blacks, whites, duckwings and piles have been seen.

In both sexes and all colours: Except that the comb, face, jaw and throat are red, there is no fixed colour for beak or legs although the beak is generally seen in yellow or ivory colour and the legs willow, white or dark olive.

Weights

Male   1.80–2.70 kg (4–6 lb)
Female 1.35–2.25 kg (3–5 lb)

Scale of points

Head (skull and beak 10, eyes 5, comb 5)	20
Neck	10
Wings	5
Thighs, shanks and feet	15
Body shape and stern	15
Plumage	10
Carriage	15
Condition	10
	100

Serious defects

Any evidence of alien blood e.g. red or dark eyes, red markings on the side of the shanks, etc. Round shanks. Duck feet. High tail carriage. Wry tail. Roach back. Stork legged or in-kneed. Any other deformity.

Bantams

Bantams should be exact miniatures of their large fowl counterparts and so standard, colour and scale of points apply.

Weights

Male   1130 g (40 oz)
Female 910 g (32 oz)

Australorp

Large fowl

Origin: Great Britain
Classification: Heavy
Egg colour: Tinted to brown

The claim that the Australorp – an abbreviation of Australian black Orpington – is the prototype of the black Orpington, as originally made by Mr. W. Cook, has never been questioned. Its breeders emphasize that its true utility type gives to poultrymen the Orpington at its best, an excellent layer and a good table fowl, with white skin.

It was around 1921 that large importations of stock birds were made from Australia into this country and an Austral Orpington Club founded. Later the breed name of Australorp was adopted, and this remains today.

General characteristics: male

Carriage: Erect and graceful, denoting an active fowl, the head being carried well above the tail line.

Type: Body deep and broad, showing somewhat greater length than depth. Back broad across shoulders and the saddle, with a sweeping curve from neck to tail. Breast full and rounded, carried well forward without bulging; breast bone long and straight. Wings compact and carried closely in, the ends being covered by the saddle hackles. Tail full and compact, rising gradually from the saddle in an unbroken line; the sickles gracefully curved, but not long and streaming.

Head: Finely modelled with skull rounded. Beak slightly curved, strong, of medium length. Eyes, large, prominent and expressive; high in skull standing out well when viewed from front or back. Comb single, medium in size, erect, evenly serrated (four to six serrations) and blade tending downwards without touching the neck, texture fine, but not of polished appearance. Face full, fine in texture, clean, free from feathers, wrinkles and overhanging brows. Ear-lobes small and elongated. Wattles medium in size, rounded at bottom and corresponding in texture to comb.

Neck: Fairly long, fine at the junction of head, with a gradual outward curve to the back, widening distinctly at the shoulders.

Legs and feet: Legs medium in length, strong, rounded in front and spaced well apart, the hocks nearly covered by body feathering, and the whole of the shanks showing below the underline. Shanks and feet (four toes) free from feathers or down.

Plumage: Feathering soft but close, with a minimum of fluff, the lower body fluff only sufficient to cover the thighs.

Skin: Fine in texture.

Australorp

Australorp large fowl
Male
Female

Australorp

Female

The general characteristics are similar to those of the male, allowing for the natural sexual differences. The pelvic bones should be pliable, not showing an excess of fat or gristle; the abdominal skin being pliable without an excess of internal fat. All these parts to be of fine texture; any indication of coarseness should be discountenanced.

Colour

Male and female plumage: Black with lustrous green sheen.

In both sexes: Beak black. Eyes black or dark brown iris, black preferred. Face, comb, ear-lobes and wattles bright red. Legs and feet black with white soles. Skin white.

Weights

Cock 3.85–4.55 kg (8½–10 lb); cockerel 3.40–4.10 kg (7½–9 lb)
Hen 2.95–3.60 kg (6½–8 lb); pullet 2.50–3.20 kg (5½–7 lb)

Scale of points

Type	35
Head (eyes 10, face 5, skull 5, comb and wattles, 5)	25
Plumage (colour, quality and character of feathering)	12
Texture and freedom from coarseness	15
Condition	8
Legs and feet	5
	100

Serious defects

Red, yellow or white in feathers. Permanent white in ear-lobes.

Defects (for which birds should be passed)

Any deformity such as wry tail, roach back, crooked breast bone, crooked toes, webbed feet. Yellow or willow colour in legs or feet. Yellow or pearl coloured eyes. Feathering on shanks or feet. Side sprigs on comb. Split or twisted wing and slipped wing.

Bantams

Australorp bantams are to be exact miniatures of their large fowl counterparts and so standard, colour and scale of points to apply.

In addition a blue Australorp has been admitted to the bantam standard. Type, carriage, head, neck, legs, feet plus quality and type of plumage to be exactly as for blacks. Weights also to be as for blacks.

Colour

The blue
Male plumage: Hackles, saddle, wing bow, back and tail a uniform dark slate blue. Remainder medium slate blue, each feather to show a wide band of lacing of a darker shade.
Female plumage: Head and neck dark slate blue, remainder medium slate blue, laced with a darker shade.

Undercolour to tone with surface colour in both sexes.

In both sexes: Beak blue or black, black preferred. Eyes black or very dark brown, black preferred. Comb, face, wattles and ear-lobes bright red. Legs and feet black or blue. Toe-nails preferably white. Skin white. Soles of feet white.

Weights

Male   1020 g (36 oz) max.
Female   790 g (28 oz) max.

Scale of points

Type and carriage	30
Colour and plumage	30
Head (eyes 10, face, comb, wattles 10)	20
Size and condition	10
Legs and feet	10
	100

Points to aim for: Clearness of blue with distinct lacing; medium length of leg; well-serrated combs.

Serious defects

As for blacks.

Autosexing breeds

An autosexing breed is one in which the chicks at hatching can be sexed by their down colouring. It was when crossing the gold Campine with the barred Rock in 1929 that Professor R. C. Punnett and Mr. M. S. Pease discovered the basic principle in their experimental work at Cambridge, and made the Cambar.

Barring is sex-linked, there being a double dose in the male and a single dose in the female, the barring being indicated by the light patch on the head of the chick. This light patch is very similar in chicks of both sexes having black down, but when the barring is transferred to a brown down there is a marked difference. The light head-spot on the female chick (one dose) is small and defined, while on the male chick (double dose) it spreads over the body. For that reason, the down colouring in the day-old cockerel is much paler, and the pattern of markings more blurred, than in the newly-hatched pullet chick, which has the sharper pattern of markings.

Standards which have been accepted by the Poultry Club are gold and silver Brussbar; Brockbar; gold, silver and cream Legbar; gold and silver Cambar; gold and silver Dorbar; Rhodebar, silver Welbar and Wybar. No bantam autosexing breeds are standardized. The Wybar standard is published below and all other standards for autosexing breeds are held by the Rare Breeds Society.

Wybar

Origin: Great Britain
Classification: Heavy
Egg colour: Tinted

Like all other autosexing breeds, the Wybar was first launched at the Cambridge School of Agriculture in 1941. Breeds used in its make-up were the light Sussex, Brussbar (barred brown Sussex), Canadian barred Rocks, and later on, Rhode Island Reds. The main Wyandotte variety used was the silver laced. The outcome was a large all-round bird suitable for both eggs and meat.

General characteristics: male

Type: Body short and deep, well rounded at sides. Back short, with broad full saddle to tail with concave sweep. Breast full round, with a straight keel. Wings medium size closely folded to the side. Tail full but short, well developed and spread at base, the true tail feathers carried rather upright, sickles medium length and gracefully curled.

Head: Broad and short, beak stout and curved. Comb rose firm, square and low in front, tapering evenly towards the back and ending in a well defined leader following the curve of the neck. Top of comb evenly covered with small rounded joints. Face smooth and fine. Ear-lobes oblong. Wattles medium, fine and rounded.

Neck: Well arched, medium length, with full hackle.

Legs and feet: Legs of medium length, thighs well covered with soft feathers, the fluff abundant, but close and silky. Shanks strong, well rounded and free from feathers. Toes, four, straight, and well covered.

Plumage: Fairly close and silky.

Female

The general characteristics are similar to those of the male allowing for natural sexual differences.

Colour

The silver
Male plumage: Evenly pale slaty cuckoo on breast, belly and shoulders; very silvery cuckoo on neck and saddle hackles, on wing coverts; tail sickles even slaty cuckoo. Gold feathers to disqualify.
Female plumage: Even silver laced (but rather pale) on breast, back, saddle, and wing coverts; tail feathers and wing primaries silvery slaty and peppered; neck hackles pale silvery cuckoo; belly and thighs faintly (but certainly) cuckoo.
Male down colour: Rather pale silvery slaty with diffuse light head patch.
Female down colour: Dark silvery slaty, or dark silver grey stripe, mottled over head: light head patch, faint or very restricted in size.

The gold
Male plumage: Evenly slaty cuckoo on breast, mahogany tinged; belly and tail feathers evenly slaty cuckoo. Wing coverts rich mahogany gold, neck and saddle hackles pale gold barred. Wings indefinitely barred. Primaries, gold-barred outer web and cuckoo inner web. Secondaries similar but reversed.
Female plumage: Evenly dark gold laced (chestnut or mahogany tinged) on breast, back, saddle, and wing coverts. Tail and wing primaries slaty chocolate, gold peppered; belly and thighs faintly (but certainly) cuckoo.
Male down colour: Rather pale smoky gold, with diffuse light head patch.
Female down colour: Dark smoky gold, or dark brown stripe, mottled over head: light head patch faint, or very restricted in size.

Weights

Cock 3.60–4.10 kg (8–9 lb); cockerel 2.90–3.40 kg (6½–7½ lb)
Hen 3.20 kg (7 lb); pullet 2.50 kg (5½ lb)

Barnevelder

Large fowl

Origin: Holland
Classification: Heavy
Egg colour: Brown

This breed was originated in the district of Barneveld, Holland, and stock was imported into this country about 1921, with the brown egg as one of the chief attractions. At first the birds were very mixed for markings, some being double laced, others single, while the majority followed a partridge or 'stippled' pattern. Two varieties were standardized, namely, double laced and partridge or 'stippled', but the former gradually came to the top, and is the popular variety of today.

Barnevelder

General characteristics: male

Carriage: Alert, upright and well balanced, the body appearing compressed, and the back concave.

Type: Body of medium length, deep and broad, with broad shoulders and high-set saddle. Breast and rump deep, broad and full. Wings rather short and carried high. Tail full, with graceful and uniform sweep.

Head: Carried high with neat skull. Beak short and full. Eyes very bold, bright and prominent. Comb single, upright, of medium size and well serrated, with a firm base, the heel to follow the neck. Face smooth, and as free from feathers as possible. Ear-lobes long. Wattles of medium size.

Neck: Fairly long, full and carried erect.

Legs and feet: Thighs and shanks of medium length to give symmetry. Shanks and feet free from feathers. Toes, four, well spread.

Plumage: Fairly tight and of nice texture.

Female

The general characteristics are similar to those of the male, allowing for the natural sexual differences.

Colour

The black
Male and female plumage: Black with beetle green sheen. This variety is now comparatively extinct.

The double laced
Male plumage: Neck and saddle hackles to match for colour and definition, each feather to be black (beetle green) with slight red-brown edging and red-brown centre quill (stem) finishing black to tip. Breast red-brown with black (beetle green) outer edging or lacing. Back and cape red-brown feathers with very wide black lacing. Abdomen and thighs black (beetle green) with black down. Wing bow and bar red-brown with broad lacing; primaries, inner edge black, outer red-brown; secondaries, inner edge black, outer red-brown finely laced with black, showing when closed as a red-brown bay. Tail, all main feathers black, with beetle green sickles and hangers. All visible black feathers and lacing to show beetle green sheen. Undercolour dark slate.

Female plumage: Hackle black with beetle green sheen. Breast, saddle, back and thighs red-brown ground clear of peppering, each feather with defined glossy black outer lacing, and inner defined lacing, the outer lacing to be distinct yet not so heavy as to give a black appearance to the bird in the show-pen. Abdomen black with black down preferred. Wing primaries inner edge black, outer brown; when wing is closed a brown bay is formed; secondaries inner edge black, outer brown, finely laced with black. Tail, main feathers black with laced feathers well up to them; undercolour grey.

Barnevelder

Barnevelder large fowl
Male
Female

Barnevelder

The partridge
Male plumage: Neck and saddle hackles red-brown with distinct but small black tip; fluff grey; quill red-brown. Breast black (beetle green). Abdomen and thighs black (beetle green) with black down. Back, cape and wing bow red-brown with wide black tip; fluff grey; quill red-brown. Wing bar black; bay brown; primaries, inner edge black, outer brown; secondaries, inner edge black, outer brown (seen as wing is closed). Tail, main feathers black with beetle green sheen; coverts, upper black, lower red-brown peppered with black; sickles black with beetle green sheen. All visible black feathers with beetle green sheen.

Female plumage: Hackle black with beetle green sheen. Breast, saddle, back and thighs red-brown ground evenly stippled with small black peppering, clear of defined inner lacing or pencilling, each feather with glossy black outer lacing, not so broad as to make the bird appear black when seen in the show-pen. Wing primaries, inner edge black, outer brown peppered with black; secondaries, outer edge brown evenly stippled with small black peppering. Tail, main feathers black, coverts peppered; undercolour grey.

The silver
Male plumage: Hackle silver with black centres. Breast silver with black edging. Back and saddle black centre with white edges. Undercolour silver grey. Wing primaries black; secondaries black edged with white. Tail black with beetle green sheen; sickles edged with white.

Female plumage: Hackle black centre with white edges, a little rust permissible. Breast white, slightly peppered, outside edge black. Wing primaries black inside, white outside, slightly peppered; secondaries well peppered.

In both sexes and all colours: Beak yellow with dark point (in the silver, horn). Eyes orange. Comb, face, wattles and ear-lobes red. Legs and feet yellow.

Weights
Cock 3.20–3.60 kg (7–8 lb); cockerel 2.70–3.20 kg (6–7 lb)
Hen 2.70–3.20 kg (6–7 lb); pullet 2.25–2.70 kg (5–6 lb)

Scale of points

Type and colour	30
Colour	25
Texture	15
Head	10
Legs and feet	10
Health and condition	10
	100

Serious defects
White in lobes. Squirrel or wry tail. Feathered legs or toes. Side sprigs on comb. Crooked toes. High or roached back. Seriously deformed breast bones. More than four toes on either foot. Black legs.

Minor defects

White in undercolour, flights, tails, wings, sickles or fluff.

Bantams

Barnevelder bantams are exact replicas of their large fowl counterparts and so standard, defects and scale of points apply. However, silvers are not standardized in bantams.

Weights

Male   910 g (32 oz) max.
Female 740 g (26 oz) max.

Belgian Bearded Bantams

The only varieties of Belgian bantams yet standardized in Britain are Barbu d'Anvers (Bearded Antwerp) and Barbu d'Uccle (Bearded Uccle). The following standards are adopted by the British Belgian Bantam Club from Belgian National Standards.

Belgian Bearded bantams are old-established true bantams, without counterparts in large breeds. Each of the two main varieties (d'Anvers and d'Uccle) has many colour variations, some of them intricate, and all attractive.

Barbu d'Anvers

General characteristics: male

(The d'Anvers is always rose combed and clean legged.)

Carriage and appearance: Small, proud, standing bold upright, with the head thrown well back; proud and provoking (appearing always ready to crow) with characteristic great development of neck hackle.

Type: Body broad and short, with arched breast carried well up. Back very short, slanting downwards to tail. Wings medium length, carried sloping towards ground. Tail carried almost perpendicularly, the main tail feathers strong and not hidden by the narrow sickle feathers; the two largest sickles slightly curved and sword-shaped, the remainder in fan-like tiers to junction with saddle hackle.

**Belgian bantam
Blue Quail D'Anvers male**

Head: Appearing rather large. Beak short, strong and curved, carrying a longitudinal band of light or dark colour in keeping with the plumage. Comb curved, broad in front, ending in a leader or spike at rear; for preference covered with small tooth-like points, or alternatively hollowed and ridged. Point or leader to follow line of neck. Eyes large and prominent, as dark as possible, colour to vary in keeping with plumage. Face covered with relatively long feathers, standing away from the head, sloping backwards and forming whiskers which cover ears and ear-lobes. Brow heavily furnished with feathers. Beard composed of feathers turned horizontally backwards from both sides of the beak and from the centre vertically downwards, the whole forming a trilobe effect. Ear-lobes small, wattles rudimentary only, but preferably none.

Neck: Of moderate length, the hackles thick and convexly arched, entirely covering back and base of neck and forming closely-joined cape at front.

Legs and feet: Thighs short, with medium-length shanks free from feathers. Toes, four, strong and straight, with nails of same colour as the beak.

Female

With certain exceptions the general characteristics are similar to those of the male, allowing for the natural sexual differences.

Carriage and appearance: A little bird, compact, plump, very lively, with characteristically full rounded neck hackle and well-developed whiskers.

Tail: Short, carried sloping upwards, towards the end and a little open.

Head: Appearing broader than that of the male and more owl-like.

Neck: Hackle inclining backwards and forming a ruffle behind the neck, with feathers broader and more developed than in the male. The female hackle, contrary to that of the male, diminishes in thickness towards bottom of neck.

Weights

As small as possible. The British Belgian Bantam Club does not advocate a weight standard for the breed, but purely as a general guide, suggests with usual variations for age and maturity:
Male   680–790 g (24–28 oz) max.
Female 570–680 g (20–24 oz) max.

Scale of points

Type (muff and beard 15, neck hackle 15, wings and tail 10)	40
Head, comb and beak	10
Colour	15
Size	15
Legs and feet	5
General appearance	15
	100

Serious defects

Wattles strongly developed. Conspicuous ear-lobes. Squirrel or wry tail. Excessive length of leg.

Disqualifications

Any trace of faking. Wattles cut or removed. Single comb. Absence of beard or whiskers. Feathers on shanks or feet. More than four toes. Yellow colouring of legs, feet or skin.

Barbu d'Uccle

General characteristics: male

(The d'Uccle is always single combed and feather legged.)

Carriage and appearance: Typically male with a majestic manner, short and broad, with characteristic heavy development of plumage.

Type: Body broad and deep. Back very broad, almost hidden by enormous neck hackle. Breast extremely broad, the upper part very developed and carried forward, the lower part resembling a breast plate. Wings close, fitting tight to body, sloping downwards and incurved towards but not beyond the abdomen; wing butts covered by neck hackle and tips (or ends of flights) covered by saddle hackle, which should be abundant and long. Tail well furnished, close and carried almost perpendicularly to line of back, the two main sickles slightly curved, the remainder in regular tiers and fan-like down to junction with saddle hackle.

Head: Slender and small, with a longitudinal depression towards the neck. Beak short and slightly curved. Comb single, fine, upright, less than average size, evenly serrated, rounded in outline, blade following line of neck. Eyes round, surrounded with bare skin. Brow heavily covered with feathers becoming gradually longer towards the rear, with a tendency to join behind the neck. Beard as full and developed as possible, composed of long feathers turned horizontally from the two sides of beak, and vertically under the beak downwards, the whole forming three ovals in a triangular group. Ear-lobes inconspicuous. Wattles as small as possible.

Neck: Furnished with silky feathers starting behind the beard at sides of throat, with a tendency to join behind the neck to form a mane. Hackle very thick and convexly arched, reaching to shoulder and saddle and covering the whole back.

Legs and feet: Legs strong, short and well apart, the hocks having clusters of long stiff feathers close together, starting from the lower outer thigh, inclined downwards and following outline of wings. Front and outside of shanks must be covered with feathers, short at top of shanks and gradually increasing in length towards the foot feather; footings turned outwards horizontally, with ends slightly curved backwards. Outer toe and outside of middle toe covered with feathers similar to shank feather.

Female

With certain exceptions the general characteristics are similar to those of the male, allowing for the natural sexual differences.

Carriage and appearance: A quiet little bird, short, thick and cobby.

Tail: Short, flat in width and not high, the lower main feathers diminishing evenly in length.

Beard: Resembling that of the male but formed with softer and more open feathers.

Neck: Hackles very thick and convexly arched, composed of broad and rounded feathers, the shape of the mane resembling that of the male.

Weights

Dwarf, as small as possible. The British Belgian Bantam Club does not advocate a weight standard, but purely as a general guide, suggests, with usual variations for age and maturity:
Male   790–910 g (28–32 oz) max.
Female 680–790 g (24–28 oz) max.

Scale of points

Type (muff and beard 15, neck hackle 15, feet and hocks 15, wings and tail 10)	55
Head, comb and beak	10
Colour	15
Size	5
General appearance	15
	100

Serious defects

Strongly developed wattles. Conspicuous ear-lobes. Squirrel or wry tail. Excessive length of leg.

Disqualifications

Any trace of faking. Wattles cut or removed. Comb other than single. Absence of beard or whiskers. Poorly feathered shanks or feet. More than four toes. Yellow legs, feet or skin.

Colour

Main colours only are fully described. Belgian bantams exist in an extraordinary choice of colours, probably unequalled in any other breed, and much too numerous to be given in detail.

Comb, ear-lobes and rudimentary wattles are red in all colour-varieties.

The millefleur
Male plumage: This is a very intricate and attractive colour scheme. Briefly the head is orange-red with white points. The beard is of black feathers laced with very light chamois, each feather ending with a round black spot with a white triangular tip. Neck hackle black with golden shafts, and broadly bordered with orange-red, each feather having a black end tipped with a white point. The extraordinary abundance of neck hackle makes the main colour appear wholly orange-red, the black parts being scarcely visible. Back red, shading to orange towards saddle hackle. Wing bows mahogany red, each feather tipped with white. Wing bars russet red with lustrous green-black pea-shaped spots at ends, finishing with silvery white triangular tips, the whole forming regular bars across the wings.

Barbu d'Uccle

Belgian bantam
Barbu D'Uccle
millefleurs female

Primaries black with a thin edging of chamois on outside. The visible lower third of each secondary feather chamois, upper two-thirds black. Remainder of wing a uniform chamois, each feather having at end a large pea-shaped white spot on a black triangle, the tips spaced evenly to conform with shape and outline of wing. (Note the reversal of these pattern-markings from the normal arrangement.) Tail feathers black with a metallic green lustre, having a fine edging or lacing of dark chamois, and terminating with a white triangle. Breast, foot feathering and remainder of plumage throughout of golden chamois ground colour, each feather having a light chamois shaft and finished with a black pea-shaped spot tipped with a white triangle.

Female plumage: Ground colour uniform golden chamois, each feather terminating with a black pea-shaped spot tipped with a white triangle. Tail feathers black, finely laced with chamois and with white tips. Wing markings and other plumage as described for male, allowing for natural sexual differences.

In both sexes: Eyes orange-red with black pupils. Beak and nails slate blue. Legs and feet slate blue.

Defects to be avoided: Ground colour too light or washed out. White markings excessively gay or unevenly distributed. Bareness or very poor quality feathering across the wing bows of the male.

The porcelaine
Male and female plumage: This is an extraordinarily delicate colour pattern. Markings and patterns generally are as described for millefleur in

Barbu d'Uccle

both sexes, with the exception that ground colour is light straw and the pea-shaped spots are pale blue, tipped with white triangles. Pale blue is substituted for the black of the millefleur in both sexes.

In both sexes: Eyes orange-red with black pupils. Beak and nails slate blue. Legs and feet slate blue.

Defects to be avoided: Ground colour too light or washed out. White markings too gay or unevenly distributed.

The quail

Male plumage: This is a very striking colour scheme, with head feathers dark green-black, finely laced with gold. Beard golden buff or nankin, shading darker towards the eyes, where plumage is black, finely laced with gold. Neck hackle with brilliant black ground, sharply laced with buff, having a golden lustre and yellowish buff shafts. Back black ground colour with gold lacing, starting in middle of feathers and narrowing towards the tips, forming lance-like points with golden silky barbs and well-defined light ochre-coloured shafts from root to point. These feathers are relatively broad under the neck, but narrower and longer nearing the saddle hackle. Colour more intense and black ground more pronounced towards the saddle hackle. Wing bows light gold, lower half of each feather black and clearly defined from the upper half, which should be nankin. Wing bars light ochre, each feather having black triangular tip, the triangles forming two regular bars across the wing. Bottom third of secondaries chamois colour, other two-thirds dull black. Primaries dull black, hidden when the wing is closed. Tail black with metallic green lustre, finely bordered with brown and with faintly-defined light shafts. Sickles black, side hangers black, laced with chamois, with well-defined light shafts. Breast nankin, each feather finely laced with ochre (yellowish buff), the shafts being distinct and clear. Thighs same colour as breast, abdomen and underparts greyish brown, with silky, golden barb-shaped tips.

The general effect is that in this variety all the upper parts are dark and the lower parts light, giving the appearance of being covered with a dark chequered cloak. The dominating dark tint is chocolate-black, with a soft silvery lustre, known amongst artists as 'umber'. The general light tone is nankin or yellow ochre, and well-defined light shafts are important.

Female plumage: Head, face and neck covered with feathers which increase in size as they near the body, ground colour umber with very fine gold lacing. Neck velvety, darker than the back and clearly detached from it. Shaft and lacing clearer and more golden towards the breast. Back covered with umber-coloured feathers having a silvery velvety lustre, each feather dark, finely laced with chamois and with bright nankin shafts showing in strong contrast. Wings the same colour as back, dark umber finely laced with chamois, feathers broader and brighter towards lower part of wing. Primaries are hidden when wing is closed and are dark intense umber. Tail plumage and cushion similar to back and of same character. Breast, clear even nankin, the shafts pale and distinct, feathers nearing the wings finely and progressively bordered with dark umber, forming a distinctive colour-pattern.

In both sexes: Eyes dark brown (nearly black) with black pupils. Legs and feet slate grey. Beak and nails horn coloured.

Defects to be avoided: Salmon or brownish colour on breast.

Barbu d'Uccle

The blue quail
Male and female plumage: Similar to the quail in all respects except that black markings are replaced by blue.

The cuckoo
Male and female plumage: Uniformly cuckoo coloured, with transverse bars of dark bluish grey on light grey ground. Each feather must have at least three bars.

In both sexes: Eyes orange-red; legs, feet, beak and nails white, often spotted with bluish grey in young birds.
Defects to be avoided: Feathers white or spotted with white, excessive number of black feathers, red on shoulders, wings and hackle.

The black mottled
Male and female plumage: All feathers black with green metallic lustre, regularly tipped with white, tips varying in size with the feather. Excessive white markings or uneven distribution to be avoided.

In both sexes: Eyes dark red; legs and feet slate blue or blackish; beak and nails dark horn.

The black
Male and female plumage: Black all over with metallic green lustre, avoiding false colouring.

In both sexes: Eyes black; legs and feet blue (blackish in young birds); beak and nails black or very dark horn.

The white
Male and female plumage: Clear white throughout, avoiding false colours, straw tinge or yellow tint on back.

In both sexes: Eyes orange-red; legs, feet, beak and nails white.

The lavender, or Reynold's blue
Male and female plumage: This is a true-breeding pale silvery blue, all the feathers being of one uniform shade.

In both sexes: Eyes orange-red with black pupils. Beak and nails slate blue. Legs and feet slate blue.
Defects to be avoided: Any straw colouring in the hackles of the male.

Other colours: These include laced blue (Andalusian type – a diffusion of black and white; blue mottled (similarly marked to black mottled); ermines (black-pointed whites); fawn ermines (black-pointed fawns or pale buffs); partridges; silvers and golds.

Not all these colours are regularly seen in this country, but there is practically no limit to the sub-varieties capable of being produced in these very charming breeds.

There are also other Bearded Belgian bantams which are not yet standardized in this country. Of these the most noteworthy is the Barbu de Watermael, which has clean legs, a small 'flying' crest or tassel, trilobed beard, and neat rose comb with three separate and distinct leaders or spikes.

Booted Bantams

Origin: Europe
Classification: True bantam
Egg colour: Tinted

The Booted Bantam or Sabelpoot as the breed is sometimes called is an ancient breed. At one time it was more popular than the present day Belgian d'Uccle bantam which it resembles. The main difference between the breeds is that there is no muffling whatever showing large round wattles and having a narrower neck characteristic.

The breed is a pure bantam with no large fowl counterpart.

General characteristics: male

Carriage: Erect and strutting.

Type: Body short and compact. Full and prominent breast. Short back, the male's furnished with long and abundant saddle feathers. Large, long wings, carried in a drooping fashion. Large tail, full and upright, the male's sickles a little longer than the main feathers and slightly curved, coverts long, abundant, and nicely curved.

Head: Skull small. Beak rather stout, of medium length. Eyes bright and prominent. Comb single, small, firm, perfectly straight and upright, well serrated. Face of fine texture, free from hairs. Ear-lobes small and flat. Wattles small, fine, and well rounded.

Neck: Rather short, with full hackle.

Legs and feet: Short. Thighs well feathered at the hocks. Fairly short shanks heavily furnished on the outer sides with long and rather stiff feathers, those growing from the hocks almost touching the ground. Toes, four, straight and well spread, the outer and the middle being very heavily feathered.

Plumage: Long and abundant.

Female

The general characteristics are similar to those of the male, allowing for the natural sexual differences.

Colour

The black
Male and female plumage: Black, as lustrous as possible.
In both sexes: Beak black or horn. Eyes dark red or very dark brown. Comb, face, wattles and ear-lobes bright red. Legs and feet black.

The white
Male and female plumage: Pure snow white.
In both sexes: Beak white. Eyes red. Comb, face, wattles and ear-lobes bright red. Legs and feet white.

Booted Bantams

The black mottled, millefleur and porcelaine: Colour as for Belgian bantams.

Weights

As for Belgian bantams.

Scale of points

Colour (of plumage 20, legs and beak 10)	30
Leg and foot feathering	15
Type	15
Head	15
Size	15
Condition	10
	100

Serious defects

Other than single comb, or four toes on each foot. Any deformity.

Brahma

Large fowl

Origin: Asia
Classification: Heavy
Egg colour: Tinted

The origin of the Brahma is wrapped in a deal of mystery, although it is accepted as of Asiatic ancestry. Its original name of Brahma-Pootra, after the river in India of that name, supports the first arrival of such stock in New York in 1846 from India. Stock reached this country in 1853. Much of the confusion at the time may have been associated with the lack of uniformity and of breed characteristics in the extra large and profusely feathered breeds being shipped from China. Both light and pencilled Brahmas were included in the Poultry Club's first Book of Standards issued in 1865, and the breed was developed with the pea comb as a characteristic.

General characteristics: male

Carriage: Sedate, but fairly active.

Type: Body broad, square and deep. Back short, either flat or slightly hollow between the shoulders, the saddle rising halfway between the hackle and the tail until it reaches the tail coverts. Breast full, with horizontal keel. Wings medium sized with lower line horizontal, free from twisted or slipped feathers, well tucked under the saddle feathers, which should be of ample length. Tail of medium length, rising from the line of the saddle and carried nearly upright, the quill feathers well spread, the coverts broad and abundant, well curved, and almost covering the quill feathers.

Head: Small, rather short, of medium breadth, and with slight prominence over the eyes. Beak short and strong. Eyes large, prominent. Comb triple or 'pea', small, closely fitting and drooping behind. Face smooth, free from feathers or hairs. Ear-lobes long and fine, free from feathers. Wattles small, fine and rounded, free from feathers.

Neck: Long, covered with hackle feathers that reach well down to the shoulders, a depression being apparent at the back between the head feathers and the upper hackle.

Legs and feet: Legs moderately long, powerful, well apart, and feathered. Thighs large and covered in front by the lower breast feathers. Fluff soft, abundant, covering the hind parts, and standing out behind the thighs. Hocks amply covered with soft rounded feathers, or with quill feathers provided they are accompanied with proportionately heavy shank and foot feathering. Shank feather profuse, standing well out from legs and toes, extending under the hock feathers and to the extremity of the middle and outer toes, profuse leg and foot feather without vulture hock being desirable. Toes, four, straight and spreading.

Plumage: Profuse, but hard and close compared with the Cochin.

Brahma

Female

With the exception of the neck and legs, which are rather short, the general characteristics are similar to those of the male, allowing for the natural sexual differences.

Colour

The dark
Male plumage: Head silver white. Neck and saddle hackles silver white, with a sharp stripe of brilliant black in the centre of each feather tapering to a point near its extremity and free from white shaft. Breast, underpart of body, thighs and fluff intense glossy black. Back silver white, except between the shoulders where the feathers are glossy black laced with white. Wing bows silver white; primaries black, mixed with occasional feathers having a narrow white outside edge; secondaries, part of outer web (forming 'bay') white, remainder ('butt') black; coverts glossy black, forming a distinct bar across the wing when folded. Tail black, or coverts laced (edged) with white. Leg feathers black, or slightly mixed with white.
Female plumage: Head silver white or striped with black or grey. Neck hackle similar to that of the male, or pencilled centres. Tail black, or edged with grey, or pencilled. Remainder any shade of clear grey finely pencilled with black or a darker shade of grey than the ground colour, following the outline of each feather, sharply defined, uniform, and numerous.

The light
Male plumage: Head and neck hackle as in the dark variety. Saddle white preferably, but white slightly striped with black in birds having very dark neck hackles. Wing primaries black or edged with white; secondaries white outside and black on part of inside web. Tail black, or edged with white. Remainder clear white, with white, blue-white, or slate undercolour, not visible when the feathers are undisturbed. Black-and-white admissible in toe feathering. Shank feathers white.
Female plumage: Neck hackle silver white striped with black (dense at the lower part of the hackle), the black centre of each feather entirely surrounded by a white margin. In other respects the colour of the female is similar to that of the male.

The white
Male and female plumage: Pure white throughout.

The gold
Male plumage: Head rich gold. Neck and saddle hackles, rich gold, with sharp stripe of brilliant black in the centre of each feather tapering to a point near its extremity and free from gold shaft. Breast, underparts of body, thighs and fluff intense glossy black. Back rich gold except between the shoulders where the feathers may be laced with gold. Wing bows bright red; primaries black, with a narrow outer edge of rich bay; secondaries, outer web (forming 'bay') partly bay, free from outer edge of black, remainder (forming 'butt') black. Wing coverts glossy black, forming a distinct bar across the wing when folded. Tail black or coverts edged with gold. Footings and leg feathers black, or slightly mixed with gold.

Brahma

Brahma large fowl
Dark male
Light female

Brahma

Female plumage: Head rich gold or striped with black. Neck hackle rich gold with sharp brilliant black striping free from shaftiness, the striping completely surrounded by gold. Tail black, edged with gold or pencilled. Remainder of plumage rich, even, clear gold, finely pencilled with black; the markings numerous, sharply defined and uniform, following the outline of the feather.

The buff columbian
Male plumage: Head golden buff. Neck and saddle hackles golden buff, with a sharp stripe of brilliant black in the centre of each feather tapering to a point near its extremity and free from buff shaft. Saddle golden buff preferably, but slightly striped with black in birds having very dark neck hackles. Wing primaries black or edged with golden buff; secondaries golden buff outside and black on part of inside web. Tail black, or edged with golden buff. Remainder a clear golden buff, with buff undercolour. Black and buff in toe feathering. Shank feather buff.

Female plumage: Neck hackle golden buff striped with black (dense at the lower part of the hackle), the black centre of each feather entirely surrounded by a golden buff margin. In other respects the colour of the female is similar to that of the male.

In both sexes and all colours: Beak yellow or yellow and black. Eyes orange-red. Comb, face, ear-lobes and wattles bright red. Legs and feet orange-yellow or yellow.

Weights
Male   4.55–5.45 kg (10–12 lb)
Female 3.20–4.10 kg (7–9 lb)

Scale of points

Type, size and carriage	35
Colour (including purity and brilliance), markings and feather	30
Legs and feet (including foot feather and leg colour)	15
Head and eye	15
Condition	5
	100

Serious defects
Comb other than 'pea' type. Badly twisted hackle or wing feathers. Total absence of leg feather. Great want of size in adults. Total want of condition. White legs. Any deformity. Buff on any part of the plumage of light. Much red or yellow in the plumage, or much white in the tail of dark males. Utter want of pencilling, or patches of brown or red in the plumage of dark females. Split or slipped wings.

Bantams

Bantams are exact miniatures of their large fowl counterparts so all standard points apply.

Weights

Male   1080 g (38 oz) max.
Female 910 g (32 oz) max.

Bresse

Origin: France
Classification: Light
Egg colour: White

The Bresse, deriving its name from the territory south of Burgundy, is a fairly firmly established favourite in France, renowned for its table qualities. There have been many attempts to popularize the breed in this country, and it is strange that a bird of such potentialities should fail to make its mark; for it is not to be despised as a layer, and it is quick maturing and hardy. As to its failure as a table bird in Britain possibly an answer is to be found in the fact that it possesses shanks of a dark slate colour, and on this side of the Channel the prejudice against skin and legs of any colour other than white is extremely strong.

General characteristics: male

Carriage;   Active and graceful.

Type: Body fairly broad and compact. Back moderately long, broad shoulders and saddle. Breast well rounded and deep. Wings long and carried close to body. Tail well developed and carried at an angle of 45° with the back; well-rounded sickle feathers.

Head: Medium sized. Beak strong and fairly short. Eyes bold. Comb single, erect and of medium size, evenly serrated, fine texture, the back part of it (the heel) clear of the neck but following the curve of the head, free from 'thumb marks' and side spikes. Face smooth and free of feathers. Ear-lobes well developed. Wattles of medium length, fine in quality and rounded at ends.

Neck: Of medium length, amply furnished with hackle feathers.

Legs and feet: Legs moderately long and well apart. Shanks free from feathers. Toes, four, straight and well spread.

Bresse

Bresse large fowl
Black male
White female

Female

With the exception of the comb, which falls gracefully over either side of the face, the general characteristics are similar to those of the male, allowing for the natural sexual differences.

Colour

The black
Male and female plumage: The plumage is black with a briliant beetle green sheen, and the undercolour black.

In both sexes: Beak dark horn. Eyes black or dark brown. Comb, face and wattles bright red. Ear-lobes snow white. Legs and feet blue-grey.

The white
Male and female plumage: Pure white, straw tinge objectionable.

In both sexes: Beak blue-white. Eyes black or dark brown. Comb and wattles bright red. Face red or sooty. Ear-lobes blue-white or white (a little red allowed). Legs and feet slate blue.

Weights

Cock 2.50–2.70 kg (5½–6 lb); cockerel 2.25–2.50 kg (5–5½ lb)
Hen 2.00–2.25 kg (4½–5 lb); pullet 1.80–2.00 kg (4–4½ lb)

Scale of points

Type	25
Head	20
Colour	15
Legs and feet	10
Size	10
Quality	10
Condition	10
	100

Serious defects

Comb with white spikes. Black, white or yellow shanks or toes. White in face. White in plumage or undercolour of the black. Straw colour or cream tinge in the white. Any deformity.

Campine

Origin: Belgium
Classification: Light
Egg colour: White

The Campine is of ancient Belgian lineage, famous for producing the finest winter milk chickens. Although not so widely kept for egg production as

Campine

formerly, it has achieved some notoriety in that it has played a major role in the work of autosex-linkage. The curiosity of the genetic workers at Cambridge was aroused by the fact that the Campine is not, as might be supposed, a barred breed like the barred Rock. The difference led to the making of the autosexing Cambar.

General Characteristics: male

Carriage: Alert and graceful.

Type: Body broad, close and compact. Back rather long, narrowing to the tail. Breast full and round. Wings large and neatly tucked. Tail carried fairly high and well spread. Campine males are hen feathered, without sickles or pointed neck and saddle hackles. The top two tail feathers slightly curved.

Head: Moderately long, deep, and inclined to width. Beak rather short. Eyes prominent. Comb single, upright, of medium size, evenly serrated, the back carried well out and clear of neck; free from excrescences. Face smooth. Ear-lobes inclined to almond shape, medium size, free from wrinkles. Wattles long and fine.

Neck: Moderately long and well covered with hackle feathers. The formation of the neck feathers in the Campine is called the cape.

Legs and feet: Legs moderately long. Shanks and feet free from feathers. Toes, four, slender and well spread.

Female

With the exception of the single comb which falls gracefully over one side of the face, the general characteristics are similar to those of the male, allowing for the natural sexual differences.

Colour

The gold
Male and female plumage: Head and neck hackle rich gold, not a washed-out yellow. Remainder beetle green barring on rich gold ground. Every feather must be barred in a transverse direction with the end gold, the bars being clear and with well-defined edges, running across the feather so as to form, as near as possible, rings round the body and three times as wide as the ground (gold) colour. On the breast and underparts of the body the barrings should be straight or slightly curved; on the back, shoulders, saddle and tail they may be of a V-shaped pattern, but preferably straight.

The silver
Male and female plumage: Head and neck hackle pure white. Remainder beetle green barring on pure white ground, the markings being as in the gold.

In both sexes and colours: Beak horn. Eyes dark brown with black pupil. Comb, face and wattles bright red. Ear-lobes white. Legs and feet leaden blue. Toe-nails horn.

Campine

Campine large fowl
Silver male
Silver female

Cochin

Weights

Male   2.70 kg (6 lb)
Female 2.25 kg (5 lb)

Scale of points

Size	10
Head (comb 5, eyes 5, lobes 5)	15
Colour (cape 12, sheen 10)	22
Tail (development and carriage)	8
Legs and feet	5
Markings	30
Condition	10
	100

Interpretation

The ideal is a bird clearly, distinctly and evenly barred all over with the sole exception of its neck hackle, which should be of the ground colour of the body. So that, taking the five main points of the bird – viz. neck hackle, top (including back, shoulders and saddle), tail, wing and breast – each is of as much importance as another; and judges are requested to bear in mind that a specimen excelling in one or two particulars but defective in others should stand no chance against one of fair average merit throughout. Special attention should be paid to size, type, and fullness of front in breeding and judging Campines.

Serious defects

Sickle feathers or pointed hackles on the males. Bars and ground colour of equal width. Ground colour pencilled. Comb at the back too near the neck. Side spikes (or sprigs) on comb. Legs other than leaden blue. White in face. Red eyes. Feather or fluff on shanks. Any deformity. Dark pigmentation in combs of females. White toe-nails. Slate blue beak. Black around the eyes.

Cochin

Origin: Asia
Classification: Heavy
Egg colour: Tinted

The Cochin, as we know it today, originally came from China in the early 1950s, where it was known as the Shanghai, and later still as the Cochin-China. The breed created a sensation in this country in poultry circles because of its immense size, and table properties. Moreover, it was an excellent layer. It was developed, however, for wealth of feather and fluff for exhibition purposes to the extent that its utility characteristics were neglected, if not made impossible, in winning types. There are no Cochin bantams.

General characteristics: male

General shape and carriage: Massive and deep. Carriage rather forward, high at stern, and dignified.

Type: Body large and deep. Back broad and very short. Saddle very broad and large with a gradual and decided rise towards the tail forming an harmonious line with it. Breast broad and full, as low down as possible. Wings small and closely clipped up, the flights being neatly and entirely tucked under the secondaries. Tail small, soft, with as little hard quill as possible and carried low or nearly flat.

Head: Small. Beak rather short, curved and very stout at base. Eyes large and fairly prominent. Comb single, upright, small perfectly straight, of fine texture, neatly arched and evenly serrated, free from excrescences. Face smooth, as free as possible from feathers or hairs. Ear-lobes sufficiently developed to hang nearly or quite as low as the wattles, which are long, thin and pendent.

Neck: Rather short, carried somewhat forward, handsomely curved, thickly furnished with hackle feathers which flow gracefully over shoulders.

Legs and feet: Thighs large and thickly covered with fluffy feathers standing out in globular form; hocks entirely covered with soft curling feathers, but as free as possible from any stiff quills (vulture hocks). Shanks short, stout in bone, plumage long, beginning just below hocks and covering front and outer sides of shanks, from which it should be outstanding, the upper part growing out from under thigh plumage and continuing into foot feathering. There should be no marked break in the outlines between the plumage of these sections. Toes, four, well spread, straight, middle and outer toes heavily feathered to ends. Slight feathering of other two toes is a good sign to breeders.

Female

With certain exceptions the general characteristics are similar to those of the male, allowing for the natural sexual differences. Comb and wattles as small as possible. The body more square than the male's and the shoulders more prominent. The back very flat, wide and short, with the cushion exceedingly broad, full and convex, rising from as far forward as possible and almost burying the tail. Wings nearly buried in abundant body feathering and the tail very small. Breast full, as low as possible. General shape is 'lumpy', massive and square. Carriage is forward, high at cushion, with a matronly appearance.

Cochin

Colour

The black
Male and female plumage: Rich black, well glossed, free from golden or reddish feathers.

In both sexes: Beak yellow, horn or black. Comb, face, ear-lobes and wattles bright red. Eyes bright red, dark red, hazel or nearly black. Legs dusky yellow or lizard.

The blue
Male plumage: Hackle, back and tail level shade of rich dark blue free from rust, sandiness or bronze. Remainder even shade of blue free from lacing on breast, thighs or fluff and free from rust, sandiness or bronze.
Female plumage: One even shade of blue free from lacing; pigeon blue preferred.

In both sexes: Beak yellow, horn or yellow slightly marked with horn. Comb, face, ear-lobes and wattles bright red. Eyes dark. Legs and feet blue with yellow tinge in pads.

The buff
Male plumage: Breast and underparts any shade of lemon buff, silver buff or cinnamon provided it is even and free from mottling. Head, hackle, back, shoulders, wings, tail and saddle may be any shade of deeper and richer colour which harmonizes well – lemon, gold, orange or cinnamon – wings to be perfectly sound in colour and free from mealiness. White in tail very objectionable.
Female plumage: Body all over any even shade, free from mottled appearance. Hackle of a deeper colour to harmonize, free from black pencilling or cloudiness, cloudy hackles being especially objectionable. Tail free from black.

In both sexes: Beak rich yellow. Comb, face, ear-lobes and wattles brilliant red. Eyes to match plumage as nearly as possible, but red eyes preferred although rare. Legs bright yellow with shade of red between the scales.

The cuckoo
Male and female plumage: Dark blue-grey bars or pencilling (across the feather) on blue-grey ground, the male's hackle free from golden or red tinge, and his tail free from black or white feathers.

In both sexes: Beak rich bright yellow, but horn permissible. Comb, face, ear-lobes and wattles as in the black. Eyes bright red. Legs brilliant yellow.

The partridge and grouse
Male plumage: Neck and saddle hackle rich bright red or orange-red, each feather with a dense black stripe. Back, shoulder coverts, and wing bow rich red, of a more decided and darker shade than the neck. Wing coverts green-black, forming a wide and sharply cut bar across the wing; secondaries rich bay outside the black inside, the end of every feather black; primaries very dark bay outside and dark inside. Saddle rich red or orange-red, the same colour as, or one shade lighter than, the neck. Remainder glossy black, as intense as possible, white in tail objectionable.

Cochin

Cochin large fowl
White male
Blue female

Crève-Coeur

Female plumage: Neck bright gold, rich gold, or orange-gold, with a broad black stripe in each feather, the marking extending well over the crown of the head. Remainder (including leg feathering) brown distinctly pencilled in crescent form with rich dark brown or black, the pencilling being perfect and solid up to the throat.

In both sexes: Beak yellow or horn. Comb, face, ear-lobes and wattles as in the black. Eyes bright red. Legs yellow, but may be of a dusky shade.

The white
Male and female plumage: Pure white, free from any straw or red shade.

In both sexes: Beak rich bright yellow. Comb, face, ear-lobes and wattles as in the black. Eyes pearl or bright red. Legs brilliant yellow.

Weights

Cock 4.55–5.90 kg (10–13 lb); cockerel 3.60–5.00 kg (8–11 lb)
Hen 4.10–5.00 kg (9–11 lb); pullet 3.20–4.10 kg (7–9 lb)

Scale of points

Feathering (cushion 8, fluff 7, tail 5, hackle 5, legs 10)	35
Colour (or markings in cuckoo or partridge)	20
Size	15
Head (ear-lobes 5)	15
Type	10
Condition	5
	100

Serious defects

Primary wing feathers twisted on their axes. Utter absence of leg feather. Badly twisted or falling comb. Legs other than yellow or dusky yellow, except in blacks and blues. Black spots in buffs. Brown mottling (if conspicuous) in partridge males, or pale breasts destitute of pencilling in partridge females. White or black feathers in cuckoos. Crooked back, wry tail, or any other deformity.

Crève-Coeur

Large fowl

Origin: France
Classification: Heavy
Egg colour: White

The Crève-Coeur is an old French breed not unlike an Houdan, but having a horn-type comb, four toes and a heavier broad square built body with bold upright carriage.

General characteristics: male

Carriage: Bold and upright.

Type: Body large, broad, and practically square. Well rounded breast. Flat back. Large well-folded wings. Full tail carried moderately high.

Head: Skull large, with a decidedly pronounced protuberance on top. Crest full and compact, round on top and not divided or split, inclined slightly backwards fully to expose the comb, not in any way obstructing the sight except from behind, composed of feathers similar to those of the hackles, and the ends almost touching the neck. Beak strong and well curved. Eyes full. Comb of the horn type (V-shaped), of moderate size, upright and against the crest, each branch smooth (free from tynes) and tapering to a point. Face muffled, the muffling full and deep, extending to the back of the eyes, hiding the lobes and the face. Ear-lobes small but not exposed. Wattles moderately long.

Neck: Long and graceful, thickly furnished with hackle feathers.

Legs and feet: Wide apart, short and the shanks free from feathers. Toes, four, straight and long.

Female

With the exception of the crest, which must be of globular shape and almost concealing the comb, the general characteristics are similar to those of the male, allowing for the natural sexual differences.

Colour

Male and female plumage: Lustrous green-black. No other colour is admissible, except a few white feathers in the crests of adults, which, however, are not desirable.

In both sexes: Beak dark horn. Eyes bright red, although black is admissible. Comb, face, wattles and ear-lobes bright red. Legs and feet black or slate blue.

Weights

Male   4.10 kg (9 lb)
Female 3.20 kg (7 lb)

Scale of points

Crest and muffling	30
Size	20
Comb	15
Colour	15
Type	10
Condition	10
	100

Serious defects

Coloured feathers. Loose crest obstructing the sight. Any deformity.

Bantams

Not seen at present but should follow the large fowl standard.

Croad Langshan

Origin: Asia
Classification: Heavy
Egg colour: Brown

The first importation of Langshans into this country was made by Major Croad, and as with other Asiatic breeds, controversy centred around it. Already there was the black Cochin and then the black Langshan, some contending both were one breed, and others that they were quite separate Chinese breeds. As developed here the breed was called the Croad Langshan after the name of the importer. In 1904 a Croad Langshan club was formed to maintain the original stamp of bird. The Modern Langshan has been developed along different lines, and in consequence the two types are shown in separate classes at shows.

General characteristics: male

Carriage: Graceful, well balanced, active and intelligent.

Type: Back of medium length, broad and flat across shoulders, the saddle well filling the angle between the back and the tail as seen in profile. In the male the back should appear shorter than in the female. Breast broad, deep and full (fuller in old bird) with long breast bone, the keel slightly rounded. Wings carried high, and well tucked up. Tail fan-shaped, well spread to right and left and carried rather high; it should be level with the head when the bird stands in position of attention; side hangers plentiful, and two sickle feathers on each side projecting some 15 cm (6 in) or more beyond rest. Abdomen capacious and resilient to touch, with fine pelvic bones. Saddle rather abundantly furnished with hackles.

Head: Carried well back, small for size of bird, full over the eyes. Beak fairly long and slightly curved. Eyes large and intelligent. Comb single, upright, straight, medium or rather small, free from side sprigs, thick and firm at the base, becoming rather thin, fine and smooth in texture, evenly serrated with five or six spikes (five preferred). Face free of feathers. Ear-lobes well developed, pendent, and fine in texture. Wattles fine in quality and rather small.

Neck: Of medium length, with full neck hackle.

Croad Langshan

Croad Langsham
large fowl
Male
Female

Legs and feet: Legs sufficiently long to give a graceful carriage to the body which should be well balanced, an adult bird neither high nor low on leg. Thighs rather short but long enough to let the hocks stand clear of fluff, well covered with soft feathers. Shanks medium length, well apart, feathered down outer sides (neither too scantily nor too heavily). Toes, four, long, straight and slender, the outer toe feathered.

Plumage: Rather soft, neither loose nor tight.

Table merits: Size for table purposes must be a great consideration, consistent with type. Bone medium or rather fine, in due proportion to size, but subordinate to amount of meat carried. Male has higher proportion of bone than female. Skin, thin and white; flesh white.

Female

Hocks need not show in adult as she carries more fluff than the male. Cushion fairly full but not obtrusive. Tail may have two feathers slightly curved and projecting about 2.5 cm (1 in) beyond rest. In other respects similar to the general characteristics of the male, allowing for the natural sexual differences.

Colour

Male and female plumage: Surface dense black with beetle green gloss free from purple or blue tinge. Undercolour dark grey, darker in the female. White in foot feather characteristic and not a defect.

In both sexes: Beak light to dark horn, preferably light at tip and streaked with grey. Eyes brown, the darker the better, but not black (ideal is colour of ripe hazel nut, a Vandyke brown). Comb, face, wattles and ear-lobes brilliant red. Shanks bluish black (bluish in adult birds, scales and toes nearly black in young birds) showing pink between scales especially on back and inner side of shank. In male bird intense red should show through the skin along outer side at base of shank feathers. Toes, the web and bottom of foot pinkish white, the deeper the pink the better; black spots on soles a serious fault. Toe-nails white; dark colour or black a serious fault.

Weights

Male 4.10 kg (9 lb) min.
Female 3.20 kg (7 lb) min.

Scale of points

Type and condition (shape 15, condition 10)	25
Body (girth 15, frame and bone 10)	25
Plumage (colour 15, furnishings and footings 10)	25
Head, feet and abdomen (head and feet 15, abdomen and pelvis 10)	25
	100

Note: It is left to discretion of judge to penalize any bad fault to the extent of 25 points.

Disqualifications

Yellow legs or feet. Yellow in face at base of beak or in edge of eyelids. Five toes. Other than single comb. Permanent white in ear-lobes. Grey (light slate colour) in webbing of flights. Black or partly black soles of feet as distinct from black spots. Vulture hocks.

Not objectionable (in stock birds and not seriously against birds in show-pen, judges using their discretion)

Purple or blue barring in few feathers only where others are of good colour. Dark red in few feathers in neck hackle or on shoulders of male. White (not grey) in flights and secondaries. White tips on head of adult female. White tips or edging on breast in chicken feathers. Moderate amount of feathering on middle toe.

Highly objectionable (to be firmly discouraged)

Appreciable amount of purple or blue barring. Decided purple or blue tinge. Light eye (make some allowance for age). Yellow iris. Wry or squirrel tail. Marked scarcity or absence of leg and foot feather.

Bantams

Standard to follow large fowl.

Weights

Male   770–910 g (27–32 oz)
Female 650–790 g (23–28 oz)

Dorking

Large fowl

Origin: Great Britain
Classification: Heavy
Egg colour: Tinted

Its purely British ancestry makes the Dorking one of the oldest of domesticated fowls in lineage. A Roman writer, who died in A.D. 47, described birds of Dorking type with five toes, and no doubt such birds were found in England by the Romans under Julius Caesar. By judicious crossings, and by careful selection, the Darking or Dorking breed was established.

Dorking

General characteristics: male

Carriage: Quiet and stately, with breast well forward.

Type: Body massive, long and deep, rectangular in shape when viewed sideways, and tightly feathered. Back broad and moderately long with full saddle inclined downward to the tail. Breast deep and well rounded with a long straight keel bone. Wings large and well tucked up. Tail full and sweeping carried well out (a squirrel tail being objectionable) with abundant side hangers and broad well-curved sickles.

Head: Large and broad. Beak stout, well proportioned and slightly curved. Eyes full. Comb single or rose. Either kind is allowed in darks, single only in reds and silver greys, and rose only in cuckoos and whites. The single comb is upright, moderately large, broad at base, evenly serrated, free from thumb marks or side spikes. The rose is moderately broad and square fronted, narrowing behind to a distinct and slightly upturned leader, the top covered with small coral-like points of even height, free from hollows. Face smooth. Ear-lobes moderately developed and hanging about one-third the depth of the wattles, which are large and long.

Neck: Rather short, covered with abundant hackle feathers falling well over the back, making it appear extremely broad at the base, and tapering rapidly at the head.

Legs and feet: Legs short and strong. Thighs large and well developed but almost hidden by the body feathering. Shanks short, moderately stout and round (square or sinewy bone being very objectionable), free from feathers, the spurs set on the inner side and pointing inwards. Toes, five, large, round and hard ('spongy' feet to be guarded against), the front toes (three) long, straight and well spread, the hind toe double and the extra toe well formed, viz. the normal toe as nearly as possible in the natural position, and the extra one placed above, starting from close to the other, but perfectly distinct and pointing upwards.

Female

The general characteristics are similar to those of the male, allowing for the natural sexual differences, except that the tail is carried rather closely. The single comb, too, falls over one side of the face.

Colour

The cuckoo
Male and female plumage: Dark grey or blue bands (barring) on light blue-grey ground, the markings uniform, the colours shading into each other so that no distinct line or separation of the colours is perceptible.

The dark
Male plumage: Hackles (neck and saddle) white or straw more or less striped with black. Back various shades of white, black and white or grey, mixed with maroon or red (bronze objectionable). Wing bows white, or white mixed with black or grey; coverts (or bar) black glossed with green;

Dorking

Dorking large fowl
Silver grey male
Dark female

secondaries outer web white, inner black. Breast and underparts jet black; white mottling not permissible. Tail richly glossed black, and a little white on primary sickles is permissible, but white hangers decidedly objectionable.

Female plumage: Neck hackle white or pale straw, striped with black or grey-black. Breast salmon red, each feather tipped with dark grey verging on black. Tail nearly black, the outer feathers slightly pencilled. Remainder of plumage nearly black, or approaching a rich dark brown, the shaft showing a cream white, each feather slightly pale on the edges, except on the wings, where the centre of the feather is brown-grey covered with a small rich marking surrounded by a thick lacing of the black, and free from red. Another successful colour is every feather over the body pencilled a brown-grey in the centre, with lacing round, and the breast as described above.

The red

Male plumage: Hackles (neck and saddle) bright glossy red. Back and wing bows dark red. Remainder of plumage jet black glossed with green.

Female plumage: Hackle bright gold heavily striped with black. Tail and primaries black or very dark brown. Remainder of plumage red-brown, the redder the better, each feather more or less tipped or spangled with black, and having a bright yellow or orange shaft.

The silver grey

Male plumage: Hackles (neck and saddle) silver white free from straw tinge or marking of any kind. Back, shoulder coverts and wing bow silver white free from striping. Wing coverts lustrous black with green or blue gloss; primaries black with a white edge on outer web; secondaries white on outer and black on inner web, with a black spot at the end of each feather, the corner of the wing when closed appearing as a bar of white with a black upper edge. Remainder of plumage deep black, free from white mottling or grizzling, although in old males a slight grizzling of the thighs is not objectionable.

Female plumage: Hackle silver white, striped with black. Breast robin red or salmon red ranging to almost fawn, shading off to ash grey on the thighs. Body clear silver grey, finely pencilled with darker grey (the pencilling following the outer line of the feather), free from red or brown tinge or black dapplings.

Note: The effect may vary from soft dull grey to bright silver grey, an old fashioned grey slate best describing the colour. Tail darker grey, inside feathers black.

The white

Male and female plumage: Snow white, free from straw tinge.

In both sexes and all colours: Beak white or horn, dark horn permissible in the dark. Eyes bright red. Comb, face, wattles and ear-lobes brilliant red. Legs and feet (including nails) a delicate white with a pink shade.

Weights

Cock 4.55–6.35 kg (10–14 lb); cockerel 3.60–5.00 kg (8–11 lb)
Hen 3.60–4.55 kg (8–10 lb)

Faverolles

Scale of points

	Dark	Silver grey or red	Cuckoo or white
Size	28	18	15
Type	20	12	20
Colour	12	24	15
Fifth toe	10	10	15
Condition	12	12	10
Head	10	16	17
Feet, condition of	8	8	8
	100	100	100

Serious defects

Total absence of fifth toe. Legs other than white or pink-white, or with any sign of feathers. Spurs outside the shank. Single comb in cuckoo or white. Rose comb in red or silver grey. White in breast or tail of silver grey male. Any coloured feathers in white. Very long legs. Crooked or much swollen toes. Bumble feet. Any deformity.

Bantams

Standards for large fowl to be used for bantams.

Weights

Male   1130–1360 g (40–48 oz)
Female 910–1130 g (32–40 oz)

Faverolles

Large fowl

Origin: France
Classification: Heavy
Egg colour: Tinted

Originated in the village of Faverolles, in Northern France, this breed was created for its dual-purpose qualities. Its make-up includes such breeds as the Dorking, Houdan and Cochin, while light Brahma blood as well as that of the Malines may be seen in some of the varieties. Imported into Great Britain in 1886, producers of table chickens crossed it freely with the Sussex, Orpington and Indian Game.

Faverolles

General characteristics: male

Carriage: Active and alert.

Type: Body deep, thick and 'cloddy'. Back fairly long, flat and 'square', i.e. very broad across shoulders and saddle. Breast broad, keel bone very deep and well forward in front, but not too rounded (a hollow breast is very objectionable). Wings small, prominent in front, carried closely. Tail moderately long, somewhat upright, and with broad feathers (flowing tail, either low or on a level with the back, is very objectionable).

Head: Broad, flat and short, free from crest. Beak short, stout. Eyes prominent. Comb single, medium size, upright, with four to six serrations, smooth and free from coarseness or any side work. Face muffled; muffling full, wide, short and solid. Ear-lobes and wattles small, of fine texture, and partly concealed by the muffles.

Neck: Short and thick, especially near the body, which it should be well let into.

Legs and feet: Legs short and stout. Thighs wide apart. Shanks straight and medium length with width between them, and sparsely feathered to the outer toe. Narrowness or any tendency to in-kneed is very objectionable. Toes, five, the front three long, straight and well spread, the outer toe sparsely feathered, the fourth toe (quite divided from the fifth) on the ground and well back, the fifth turned up the leg.

Female

The general characteristics are similar to those of the male, allowing for the natural sexual differences, with the following exceptions. Comb much smaller in proportion, back longer in proportion, neck straighter, keel bone longer and deeper, and the tail carried midway between upright and drooping.

Colour

The black
Male and female plumage: Black showing rich beetle green sheen, free from purple bars.
   In both sexes: Beak black. Eyes black or brown. Comb red. Face, wattles and ear-lobes red, partly concealed by muffling. Legs and feet black.

The blue
Male plumage: Head, muffling, hackles, back, tail and wing bows a uniform dark blue. Remainder rich blue, each feather laced a dark shade.
Female plumage: Rich uniform blue, each feather laced a dark shade.
   In both sexes: Beak, etc. as in the black except that the legs and feet may also be blue.

Faverolles

Faverolles large fowl
Salmon male
White female

Faverolles

The buff
Males and female plumage: Rich lemon buff throughout.

In both sexes: Beak horn or white. Eyes grey or hazel. Comb red. Face, wattles and ear-lobes red, partly concealed by muffling. Legs and feet white.

The ermine
Male and female plumage: Head and neck plumage each feather having a solid black centre entirely surrounded by an even white margin. Wings white with black in flights. Tail black. Remainder pure white.

In both sexes: Beak, etc. as in the buff.

The salmon
Male plumage: Beard and muff black. Hackles straw. Back, shoulders and wing bows bright cherry mahogany. Breast, thighs, underfluff, tail and shank feathering black. Wing bar black; primaries black; secondaries white outer edge, black inner edge and at tips.
Female plumage: Beard and muff creamy white. Breast, thighs and fluff cream. Remainder wheaten brown; head and neck striped with dark shade of the same colour (free from black) and wings softer and lighter than back. Primaries, secondaries and tail wheaten brown.

In both sexes: Beak, etc. as in the buff.

The white
Male and female plumage: Pure white.

In both sexes: Beak, etc. as in the buff.

Weights

Cock 3.60–4.55 kg (8–10 lb); cockerel 2.95–4.10 kg (6½–9 lb)
Hen 2.95–3.85 kg (6½–8½ lb); pullet 2.70–3.60 kg (6–8 lb)

Scale of points

Utility qualities, size and condition	25
Type	25
Colour	20
Beard and muffling	15
Formation of feet and toes	5
Foot feather	5
Comb	5
	100

Serious defects

Skin and legs other than white (except in the black and the blue). Absence of muffling. Featherless shanks and outer toes. Other than five toes on each foot. Hollow breast. White or brassiness in hackle, wing or saddle, or purple barring, or white in foot feather in the blue. Mealiness of general colour, or white in tail, wings and undercolour of the buff. Other than white legs, smuttiness on back, in the ermine. Brassiness on wings of white male. Any bodily deformity.

Bantams

Faverolles bantams to be exact replicas of their large fowl counterparts.

Weights

Male   1130–1360 g (40–48 oz)
Female 910–1130 g (32–40 oz)

Scale of points

Almost the same except that utility qualities, size and condition have 20 points and beard and muffling also 20 points.

Frizzle

Large fowl

Origin: Asia
Classification: Heavy
Egg colour: White or tinted

The Frizzle, a purely exhibition breed, is of Asiatic origin, and is notable for its quaint feather formation, each feather curling towards the head of the bird. It is more popular in bantams than in large fowls.

General characteristics: male

Carriage: Strutting and erect.

Type: Body broad and short. Breast full and rounded. Wings long. Tail rather large, erect, full but loose, with full sickles and plenty of side hangers. Lyre tails in males desirable but not obligatory.

Head: Fine. Beak short and strong. Eyes full and bright. Comb single, medium sized and upright. Face smooth. Ear-lobes and wattles moderate size.

Neck: Of medium length, abundantly frizzled.

Legs and feet: Legs of medium length. Shanks free from feathers. Toes, four, rather thin, and well spread.

Plumage: Moderately long, broad and crisp, each feather curled towards the bird's head, and the frizzling as close and abundant as possible.

Female

The general characteristics are similar to those of the male, allowing for the natural sexual differences, except that the comb is much smaller and the neck is not so abundantly frizzled.

Frizzle

Frizzle large fowl
Grey male
Red female

Frizzle

Frizzle bantams
White male
Blue female

Frizzle

Colour

Male and female plumage: Black, blue, buff or white, a pure even shade throughout in the 'self coloured' varieties; columbian as in Wyandotte; duckwing, black-red, brown-red, cuckoo, pile, and spangle as in Old English Game; red as in Rhode Island Red.

In both sexes and all colours: Beak yellow in the buff, columbian, pile, red and white varieties; white in the spangle, black-red and cuckoo; and dark willow, black or blue in other varieties. Eyes red. Comb, face, wattles and ear-lobes bright red. Legs and feet to correspond with the beak. (There are variations in leg colour, and yellow legs are frequently demanded in blacks, though not so standardized.)

Weights

Cock 3.60 kg (8 lb); cockerel 3.20 kg (7 lb)
Hen 2.70 kg (6 lb); pullet 2.25 kg (5 lb)

Scale of points

Type	25
Colour	25
Curl of feather	30
Condition	10
Weight	10
	100

Serious defects

Narrow feather. Want of curl. Long tail. Drooping comb. Other than single comb. White lobes. Deformity of any kind.

Bantams

Frizzle bantams follow the large fowl standard although a slightly different scale of points is used.

Weights

Male 680–790 g (24–28 oz)
Female 570–680 g (20–24 oz)

Scale of points

Head and comb	5
Legs and feet	5
Plumage colour	15
Size	10
Curl	25
Feather quality	20
Type and symmetry	10
Condition	10
	100

Hamburgh

Large fowl

Origin: North Europe
Classification: Light
Egg colour: White

The origin of the Hamburgh is wrapped in mystery. The spangled were bred in Yorkshire and Lancashire three hundred years ago as Pheasants and Mooneys, and there is a book reference to black Pheasants in the North of England in 1702. In its heyday the Hamburgh was a grand layer and must have played its part in the making of other laying breeds. However, its breeders directed it down purely exhibition roads, until today it is in few hands.

General characteristics: male

Carriage: Alert, bold and graceful.

Type: Body moderately long, compact, fairly wide and flat at the shoulders. Breast well rounded. Wings large and neatly tucked. Tail long and sweeping, carried well up (but avoiding 'squirrel' carriage), the sickles broad and the secondaries plentiful.

Head: Fine. Beak short, well curved. Eyes bold and full. Comb rose, medium size, firmly set, square fronted, gradually tapering to a long, finely-ended spike (or leader) in a straight line with the surface and without any downward tendency, the top level (free from hollows) and covered with small and smooth coral-like points of even height. Face smooth and free from stubby hairs. Ear-lobes smooth, round and flat (not concave or hollow), varying in size according to the variety. Wattles smooth, round and of fine texture.

Neck: Of medium length, covered with full and long feathers, which hang well over the shoulders.

Legs and feet: Legs of medium length. Thighs slender. Shanks fine and round, free of feathers. Toes, four, slender and well spread.

Female

The general characteristics are similar to those of the male, allowing for the natural sexual differences.

Colour

The black
Male and female plumage: Rich black, with a distinct green sheen from head to tail, and especially on sickle feathers and tail coverts. Any approach to bronze or purple tinge or barring to be avoided.

In both sexes: Beak black or dark horn. Eyes, comb, face and wattles red. Ear-lobes white. Legs and feet black.

Hamburgh

Hamburgh large fowl
Silver spangled male
Silver spangled female

Hamburgh

Hamburgh large fowl
Black male
Gold pencilled female

Hamburgh

The gold pencilled
Male plumage: Bright red bay or bright golden chestnut, except the tail, which is black, the sickle feathers and coverts being laced all round with a narrow strip of gold.
Female plumage: Ground colour similar to the general colour of the male, and, except on the hackle (which should be clear of all marking, if possible), each feather distinctly and evenly pencilled straight across with fine parallel lines of a rich green-black, the pencilling and the intervening colour to be the same width, while the finer and the more numerous on each feather the better.

In both sexes: Beak dark horn. Eyes, comb, face and wattles red. Ear-lobes white. Legs and feet lead blue.

The silver pencilled
Male and female plumage: Except that the ground colour, and in the male the tail lacings, are silver, this variety is similar to the gold pencilled.

The gold spangled
Male plumage: Ground colour rich bright bay or mahogany; striping, spangling, tipping and tail rich green-black. Hackles and back, each feather striped down the centre. Wing bow dagger-shaped tips at the end of each feather; bars (two), rows of large spangles, running parallel across each wing with a gentle curve, each bar distinct and separate; secondaries tipped with large round spangles, forming the 'steppings'. Breast and underparts, each feather tipped with a round spot or spangle, small near the throat, increasing in size towards the thighs, but never so large as to overlap.
Female plumage: Ground colour and spangling are similar to those of the male. Hackle, wing bars and 'steppings' as in the male. Tail coverts black, with a sharp lacing or edging of gold on each feather. Remainder, each feather tipped with a spangle, as round as possible, and never so large as to overlap, the spangling commencing high up the throat.

In both sexes: Beak, eyes, comb, face, wattles, ear-lobes, legs and feet as in the pencilled varieties.

The silver spangled
Male plumage: Ground colour pure silver; spangling and tipping rich green-black. Hackles, shoulders and back, each feather marked with small, dagger-like tips. Wing bow dagger-shaped tips, increasing in size until they merge into what is known as the third bar; bars (two) and secondaries, breast and underparts similarly marked to those of the gold spangled variety. Tail ending with bold half-moon-shaped spangles; sickles with large round spangles at the end of each feather; coverts similar, though spangles not so big.
Female plumage: Ground colour and spangling similar to those of the male. Hackle marked from the head with dagger-shaped tips, which gradually increase in width until they merge into the spangles at the bottom. Wing secondaries as in the male, bars similar to those of the gold spangled female. Tail, each feather with a half-moon-shaped spangle at the end. Coverts reaching halfway up the true tail feathers to form a row across

Hamburgh

the tail (each side) of round spangles. Remainder marked as in the gold female.

In both sexes: Beak, eyes, comb, face, wattles, ear-lobes, legs and feet as in the pencilled varieties.

Weights
Male   2.25 kg (5 lb) approx.
Female 1.80 kg (4 lb) approx.

Scale of points

The black	Male	Female
Type, style and condition	15	15
Head (comb, face and ear-lobes 15 each)	45	45
Colour (legs 5)	25	35
Tail	15	5
	100	100

The pencilled	Male	Female
Type, style and condition	10	10
Head (comb, ear-lobes and face)	25	20
Colour (including legs)	30	10
Markings (back and cushion 15, breast and thighs 15, tail 15, wings 10, neck hackle 5)	—	60
Tail markings	35	—
	100	100

The gold spangled	Male and Female
Type, style and condition	10
Head (comb 10, ear-lobes 5, face 5)	20
Colour (including legs)	10
Markings (back and saddle 15, breast and thighs 15, wings 15, neck hackle 10, tail 5)	60
	100

The silver spangled	Male	Female
Type, style and condition	10	10
Head (comb 10, ear-lobes 10, face 5)	25	—
Head (comb 10, ear-lobes 5, face 5)	—	20
Colour (including legs)	10	10
Markings (tail 15, neck hackle 10, back and saddle 10, breast and thighs 10, wings 10)	55	—
Markings (back and cushion 15, neck hackle 15, tail 10, breast and thighs 10, wings 10)	—	60
	100	100

Serious defects

White face. Single comb. Red ear-lobes. Squirrel or wry tail. Any other deformity.

Bantams

Hamburgh bantams follow the large fowl standard except that black is not standardized as a bantam colour. The scale of points used is different.

Weights

Male   680–790 g (24–28 oz)
Female 620–740 g (22–26 oz)

These weights are to be treated as maximums although the pencilled varieties are usually considerably larger.

Scale of points

Markings	60
Head, comb, face and lobes	20
Colour	10
Type, style and condition	10
	100

In pencilled males the points allotted to markings are divided between tail and colour.

Houdan

Large fowl

Origin: France
Classification: Light
Egg colour: White

Introduced into England in 1850, the Houdan is one of the oldest French breeds, taking its name from the town of Houdan, and being developed for table qualities. Developed here it was once classified as a heavy breed, but today is included in the category of light, non-sitting breeds. It is one of the few breeds carrying a fifth toe, a semi-dominant feature when crossed with other breeds.

General characteristics: male

Carriage: Bold and active.

Type: Body broad, deep and lengthy, as in the Dorking. Tail full with the sickles long and well arched.

Houdan

Houdan large fowl Female

Head: Fairly large, with a decidedly pronounced protuberance on top, and crested. Crest full and compact, round on top and not divided or 'split', composed of feathers similar to those of the hackle, inclining slightly backwards fully to expose the comb, in no way obstructing the sight except from behind. Beak rather short and stout, well curved, and with wide nostrils. Eyes bold. Comb leaf type, somewhat resembling a butterfly placed at the base of the beak, fairly small, well defined, and each side level. Face muffled; muffling large, full, compact, fitting around to the back of the eyes and almost hiding the face. Ear-lobes small, entirely concealed by muffling. Wattles small and well rounded, almost concealed by beard.

Neck: Of medium length, with abundant hackle coming well down on the back.

Legs and feet: Legs short and stout, well apart, free of feathers. Toes, five, similar to those of the Dorking.

Female

The general characteristics are similar to those of the male, allowing for the natural sexual differences, with the exception of the crest, which is full, compact and globular, not in any way obstructing the sight except from behind, and with the comb visible. Tail fairly full.

Houdan

Colour

Male and female plumage: Glossy green-black ground with pure white mottles, the mottling to be evenly distributed, except on the flights and secondaries, and in the male the sickles and tail coverts, which are irregularly edged with white. *Note:* In young Houdans black generally preponderates, but what mottling there is should be even and clear. Mottling becomes gayer with age.

In both sexes: Beak horn. Eyes red. Comb, face and wattles bright red. Ear-lobes white or tinged with pink. Legs and feet white mottled with lead blue or black.

Weights

Male 3.20–3.60 kg (7–8 lb)
Female 2.70–3.20 kg (6–7 lb)

Scale of points

	Male	Female
Type	12	10
Size	18	20
Comb	15	8
Legs and feet	10	10
Colour	15	15
Crest	12	15
Muffling	8	12
Condition	10	10
	100	100

Serious defects

Red or straw-coloured feathers. Loose crest obstructing the sight. Spur outside the shank. Feathers on shanks or toes. Other than five toes on each foot. Any deformity.

Bantams

Houdan bantams are rarely seen but should follow the large fowl standard.

Weights

Male 680–790 g (24–28 oz)
Female 620–740 g (22–26 oz)

Indian and Jubilee Indian Game

Large fowl

Origin: Great Britain
Classification: Heavy
Egg colour: Tinted

To Cornwall must go the credit for giving us the Indian Game. Breeds used in the make-up were the red Asil, black-breasted red Old English Game, and the Malay. The breed has been developed for its abundant quantity of breast meat, in which respect no other breed can equal it. When large table birds were the most popular in this country Indian Game males were chosen as mates for females of such table breeds as the Sussex, Dorking and Orpington, to produce extra large crosses. The females chosen for mating belonged to breeds possessing white flesh and shanks. Jubilee Indian Game are similar to Indians, but the lacing is white; in Indians it is black. The two varieties are often interbred.

General characteristics: male

Carriage: Upright, commanding and courageous, the back sloping downwards towards the tail. A powerful and broad bird very active, sprightly and vigorous.

Type: Body very thick and compact and very broad at the shoulders, the shoulder butts showing prominently, but the bird must not be hollow backed, the body tapering towards the tail. Back flat and broad at the shoulders, but the bird must not be flat sided. Elegance is required with substance. Breast wide, fairly deep and prominent, but well rounded. Wings short and carried closely to the body, well rounded at the points, closely tucked at ends and carried rather high in front. Tail medium length with short narrow secondary sickles and tail coverts, close and hard; carriage drooping.

Head: Rather long and thick, not so keen as in English Game, nor as thick as in the Malay; somewhat beetle browed but not nearly as much as in the Malay. Skull broad. Beak well curved and stout where set on the head giving the bird a powerful appearance. Eyes full and bold. Comb (in undubbed birds) pea type, i.e. three longitudinal ridges, the centre one being double height of those at sides, small closely set on the head. Ear-lobes and wattles smooth and of fine texture.

Neck: Of medium length and slightly arched; hackle short, barely covering base of the neck.

Legs and feet: Legs very strong and thick. Thighs round and stout, but not as long as in the Malay. Shanks short and well scaled. The length of shank must be sufficient to give the bird a 'gamey' appearance. Feet strong

Indian and Jubilee Indian Game

Indian Game
Male
Female

Indian and Jubilee Indian Game

and well spread. Toes long, strong, straight, the back toe low and nearly flat on ground; nails well shaped.

Plumage: Short, hard and close.

Handling: Flesh firm.

Female

The general characteristics are similar to those of the male, allowing for the natural sexual differences. The tail, however, which is well venetianed but close, is carried low but somewhat higher than the male's.

Colour

Male plumage: Head, neck, breast, underfluff, thighs, and tail black, with rich green glossy sheen or lustre, the base of the neck and tail hackles a little broken with bay or chestnut, which should be almost hidden by the body of the feathers. Shoulders and wing bows green glossy black or beetle green, slightly broken with bay or chestnut in the centre of the feather or shaft. Tail coverts green glossy black or beetle green slightly broken with bay or chestnut in the base of the shaft. Back feathers green glossy black or beetle green, also touched on the fine fronds at the end of the feathers with bay or chestnut which gives the sheen so much desired. When the wing is closed there is a triangular patch of bay or chestnut formed of the secondaries, which are green glossy black or beetle green on the inner, and bay or chestnut on the outer web, and which when closed show only the bay in a solid triangle. The primaries, ten in number, are curved and of a deep black, except for about 6.25 cm (2½ in) of a narrow lacing of light chestnut on the outer web.

Female plumage: The ground colour is chestnut brown, nut brown, or mahogany brown. Head, hackle and throat green glossy black or beetle green. The pointed hackle that lies under the neck feathers green glossy black, or beetle green with a bay or chestnut centre mark; the breast commencing on the lower part of the throat, expanding into double lacing on the swell of the breast, of a rich bay or chestnut, the inner or double lacing being most distinct, the belly and thighs being marked somewhat similarly and running off into a mixture of indistinct markings under the vent and swell of the thighs. The feathers of shoulders and back are somewhat smaller, enlarging towards the tail coverts and similarly marked with double lacing; the markings on wing bows and shoulders running down to the waist are most distinct of all, with the same kind of double lacing. Often in the best specimens there is an additional mark enclosing the base of the shaft of the feather and running to a point in the second or inner lacing. Tail coverts are seldom as distinctly marked, but have the same style of marking. Primary or flight feathers are black, except on inner frond or web which is a little coloured or peppered with a light chestnut. Secondaries are black on the inner web, while the outer web is in keeping with the general ground colour and is edged with a delicate lacing of green glossy black or beetle green. Wing coverts which form the bar are laced like those of the body and often a little peppered. The black lacing should

Indian and Jubilee Indian Game

be metallic green, glossy black or beetle green. This should appear embossed or raised.

In both sexes: Beak horn, yellow, or horn striped with yellow. Eyes from pearl to pale red. Face, comb, wattles and ear-lobes rich red. Legs rich orange or yellow, the deeper the better.

Weights

Male   3.60 kg (8 lb) min.
Female 2.70 kg (6 lb) min.

Scale of points

Type and colour (body and thighs 10, back, breast, wings, tail, legs, 8 each, neck 3)	53
Carriage	12
Size	10
Head (skull, eyes and brows, 3 each; beak, wattles, lobes, comb, 2 each)	17
Condition	8
	100

Note: The Indian Game fowl is in no way allied to the English Game fowl. Hence it is not recognized as a true Game bird in the Fancy; that is, unless classes are specially provided for the breed it must compete in the 'Any other variety' classes and not in those set aside for 'Game', *vide* Poultry Club Show Rules.

Defects

Crooked breasts or toes. Flat shins. Rusty hackles. Bad shape. Heavy feathering. White in hackles. Smallness of size. Long legs and thighs. Twisted hackle.

Disqualifications

Male: Crooked back, beak and legs. Wry or squirrel tail, in-knees, bent legs and flat sides. Single or Malay comb. Red hackles. Additionally in the female too light, too dark or mealy ground colour, and defective markings.

Jubilee Indian Game

General characteristics

As for Indian Game.

Colour

Male plumage: Head, neck, breast, body, underfluff, thighs and tail white. Hackle feathers to have chestnut shaftings. Clear breasts are

Indian and Jubilee Indian Game

Jubilee Indian Game
Male
Female

desirable. Wing bows and shoulders white, slightly broken with bay or chestnut. Wing primaries and secondaries white with bay markings. Triangular patch of bay or chestnut to show when wing is closed. Tail coverts white. Back white touched with bay or chestnut.

Female plumage: Ground colour chestnut brown or mahogany. Head hackle and throat white. Breast, commencing on the lower part of the throat and expanding to double lacing on the swell of the breast, mahogany laced with white. The inner, or double lacing, to be most distinct. The underparts and thighs are marked somewhat similarly and run into a mixture of indistinct markings beneath the vent and swell of the thighs. Feathers of the shoulders and back somewhat small, enlarging towards the tail coverts similarly marked with the double lacing; often in the best specimens there is an additional mark enclosing the base of the shaft of the feather and running to a point in the second or inner lacing. The tail coverts are seldom as distinctly marked, but with the same style of marking. Wing primaries, white marked on inner web with chestnut. Secondaries, white inner web, chestnut outer web, edged with white. Main tail white. Remainder chestnut ground colour throughout, double laced with white, inner lacing should be quite distinct. Underparts and thighs may be less distinctly marked and wing coverts may be peppered.

In all other respects the Indian Game standard should be followed.
In both sexes: Beak, eyes, comb and legs as described for Indians.

Bantams

These miniatures are well established but they are not recognized as Game birds in spite of their name; and although they may compete in classes specified for 'hard feather' they may not compete in classes listed for Game birds. Where no 'hard feather' or breed classes are scheduled, they must compete in the A.O.V. section. Jubilees are similar to the Indians except that where Indians are black, Jubilees are white. Both colours are frequently interbred.

Indian and Jubilee Game bantams should follow the large fowl standard, including the scale of points.

Weights

The Indian Game Club does not issue definite weight standards for bantams and weights detailed below are suggestions only. These weights are often exceeded and excess size should be penalized.

Male   1130–1360 g (40–48 oz)
Female 910–1130 g (32–40 oz)

Originally, British Bantam Association weights were below these.

Indian and Jubilee Indian Game

Indian Game bantams
Male
Female

105

Ixworth

Origin: Great Britain
Classification: Heavy
Egg colour: Tinted

Originated at the village of Ixworth in Suffolk, this breed was introduced mainly for its table qualities allied to steady egg production. It is clear from the general outline of the breed that the Indian Game played a part in its make-up. There is no bantam version of this breed.

General characteristics: male

Carriage: Alert, active and well balanced.

Type: Body deep, well rounded, fairly long but compact. Back long, flat, reasonably broad, without too prominent a slope to the tail. Breast broad, full, deep, well rounded, long and wide, low breast bone carried well forward; with unpronounced keel or keel point; well fleshed and rounded off for entire length. Wings strong, carried close, showing shoulder butts. Tail compact, of medium length and carried fairly low, the sickles close fitting.

Head: Broad and of medium length. Beak short and stout. Eyes full, prominent, keen expression, without heavy brows. Comb pea type. Face smooth and of fine texture. Ear-lobes and wattles medium size and fine texture.

Neck: Somewhat erect and of reasonable length. Hackle feathers short, close fitting and in no way excessive or loose.

Legs and feet: Legs well apart, and of reasonable length to ensure activity. Thighs well fleshed and of medium length. Shanks covered with tight scales, free from feathers. Toes, four, straight, well spread and firm stance. Bone characteristic of a first-class table bird.

Plumage: Short, silky and close fitting; fluff likewise.

Female

The general characteristics are similar to those of the male, allowing for the natural sexual differences.

Colour

Male and female plumage: White.
   In both sexes: Beak white. Eyes red or bright orange. Comb, face, wattles, and ear-lobes brilliant red. Legs, feet, skin and flesh white.

Weights

Cock 4.10 kg (9 lb); cockerel 3.60 kg (8 lb)
Hen 3.20 kg (7 lb); pullet 2.70 kg (6 lb)

Ixworth large fowl
Male

Scale of points

Table merits	40
Shape and size	20
Colour (general)	20
Head	10
Plumage and condition	10
	100

Serious defects

Coarseness. Lack of activity. Loose feathers. Any point against table values or general usefulness. Any deformity.

Japanese Bantams

Origin: Japan

True bantams of great antiquity, these are without counterparts in large breeds. They are the shortest legged of all varieties. The standard here given is that of the International Japanese Club, which was approved and adopted by the British Club in 1937.

Japanese Bantams

General characteristics: male

Carriage and appearance: Very small, low built, broad and cobby with deep full breast and full-feathered upright tail. Appearance somewhat quaint due to a very large comb, dwarfish character and waddling gait. Plumage very full and abundant.

Type: Back very short, wide, and seen from the side it forms the shape of a small letter U, the sides being formed by the neck and tail. This shape, however, is almost lost in fully feathered males. Saddle hackles rich and long. Body short, deep and broad. Breast very full, round and carried prominently forward. Wings long with the tips of the secondaries touching the ground immediately under the end of the body. Thighs very short and not visible. Tail very large and upright. The main tail feathers should rise above the level of the head about one-third of their length, spreading well and with long sword-shaped main sickles and numerous soft side hangers. The tail may touch the comb with its front feathers, but must not be set so as to lean forward at too sharp an angle.

Head: Large and broad, beak strong and well curved, eyes large. Comb single, large (the larger the better), coarse grained, erect and evenly serrated with four or five points. The blade of the comb should follow the nape of the neck. Face smooth, ear-lobes medium size, red and free from all traces of white. Wattles pendent and large.

Neck: Rather short, curving backwards and with abundant hackle feathers which should well drape the shoulders.

Legs and feet: Shanks very short, clean (free from feather), strong and sharply angled at the joints. The shanks to be so short as to be almost invisible. Toes, four, straight and well spread.

Female

The general characteristics should follow closely those described for the male regarding type. Breast should be all as described for the male. Tail well spread and rising well above the head. The main tail feathers broad, the foremost pair being slightly curved (sword-shaped). Comb large, evenly serrated and preferably erect, although falling to one side being no defect.

Colour

The black-tailed white
Male and female plumage: Body feathers white, wing primaries and secondaries should have white outer and black inner webs, the closed wings look almost white. Main tail feathers black or black with white lacing. Main sickles and side hangers black with white edging.

In both sexes: Eyes red. Legs yellow.

The black-tailed buff
Male and female plumage: The same markings as the black-tailed white, except that the white is replaced by buff.

Japanese Bantams

Japanese bantams
Black male
Wheaten female

109

Japanese Bantams

The buff columbian
Male and female plumage: Rich even buff, wing primaries and secondaries buff with black inner webs, the closed wings look almost buff. Sickles and side hangers black with buff edges. Neck hackle feathers buff with black centre down each, the hackle to be free from black edges.

In both sexes: Eyes red. Legs yellow.

The white
Male and female plumage: Pure white without sappiness.

In both sexes: Eyes red. Legs yellow.

The black
Male and female plumage: Deep full black with a green sheen.

In both sexes: Eyes black. Comb and face red. Legs yellow, black permitted on shanks but underside of feet must be yellow.

The greys
The birchen: Male and female plumage: The hackle, back, shoulder coverts and wing bows silver white. The neck, saddle hackle and saddle with narrow black striping. Remainder charcoal grey or black, the breast feathers having a narrow light grey or silver edging around each feather, thereby giving a laced appearance. In females the neck hackle light grey or silver with narrow black striping. Breast laced as for males, remainder charcoal grey, or black.

In both sexes: May have dark comb, flesh and face.

The Miller's: Male and female plumage: As birchen, but mealy on breast.

The dark: Male and female plumage: As above but breast black. Female almost all dark with striped neck hackle.

The light: Male and female plumage: As birchen without lacing on breast, clear black breast.

In both sexes and all greys: Black wing secondaries. Beak shaded with black or horn. Eyes red, iris orange or dark brown.

The mottleds
The black: Male and female plumage: All feathers should be black with white tips. The amount of white may vary, but the ideal is between 0.94 cm (⅜ in) and 1.25 cm (½ in). Tails and wings similar but more white permitted.

The blue: Male and female plumage: As above but blue instead of black.

The red: Male and female plumage: As the black but red instead of black.

In both sexes and all mottleds: Beaks to match legs, which should be yellow or willow. Eyes red or orange.

The blues
The self: Male and female plumage: All feathers blue, neck feathers may be a darker blue than the remainder.

In both sexes: Eyes orange. Legs slate or willow.

The lavender: Male and female plumage: All feathers a lavender blue to skin, even shade throughout.

In both sexes: Eyes orange or dark brown. Legs slate or blue.

The cuckoo
Male and female plumage: The feathers throughout including body, wing and tail to be generally uniformly cuckoo coloured with transverse bars of dark bluish grey on a light grey ground.

In both sexes: Beak yellow marked with black. Eyes orange or red. Legs yellow.

The red
Male and female plumage: All feathers deep red, solid to skin an even shade throughout.

In both sexes: Beak and legs yellow, both may be marked with red. Eyes red.

The tri-coloured
Male and female plumage: The colours white, black and brown or dark ochre should be as equally divided as possible on each feather.

In both sexes: Eyes red or orange. Legs yellow or willow.

The black-red
Male and female plumage: Wheaten and partridge bred.

The brown-red, blue-red, silver and golden duckwing: The latter four colours all as described for Old English Game.

The following secondary colours are permitted: ginger, blue dun, honey dun, golden hackled, furnace.

Weights
Male   510–620 g (18–22 oz)
Female 400–510 g (14–18 oz)

Scale of points

Type	55
Size	15
Condition	15
Colour	10
Leg colour	5
	100

Serious defects
Narrow build. Long legs. Long back. Wry tail. Tail carried low. Deformed comb and lopped comb on males. High wing carriage. White in lobes. Any physical deformity.

Jersey Giant

Origin: America
Classification: Heavy
Egg colour: Tinted to brown

Originated in New Jersey in about 1880, this American breed took the name of 'Giant' because of the extra heavy weights that specimens could record. Its make-up accounts for such poundage as it includes black Java,

Jersey Giant

dark Brahma, black Langshan and Indian Game. When introduced into this country it was claimed for the breed that the birds were heavier than those of any other breed, that it was adaptable for farm range, and also for providing capons. Earlier specimens were of exceptional weights.

General characteristics: male

Carriage: Bold, alert and well balanced.

Type: Body long, wide, deep and compact; smooth at sides, with long keel, smooth and moderately full fluff. Back rather long, broad, nearly horizontal, with a short sweep to the tail. Breast broad, deep and full, carried well forward. Wings medium sized, well folded, carried at the same angle as the body, the primaries and secondaries broad and overlapping in natural order when the wings are folded. Tail rather large, full, well spread, carried at an angle of 45° above the horizontal, the sickles just sufficiently long to cover the main tail feathers, the coverts moderately abundant and of medium length, the main tail feathers broad and overlapping.

Head: Rather large and broad. Beak short, stout and well curved. Eyes large, round, full and prominent. Comb single, straight, upright, rather large and of fine texture, having six well-defined and evenly-serrated points, the blade following the shape of the neck. Face smooth and fine in texture. Ear-lobes smooth and rather large, extending down one-half of the length of the wattles. Wattles of medium size and fine texture, well rounded at the lower ends.

Neck: Moderately long, full and well arched.

Legs and feet: Legs straight and set well apart. Thighs large, strong, of moderate length and well covered with feathers. Shanks strong, stout, medium length and free from feathers; scales fine; bone of good quality and proportionate to size of bird. Toes, four, of medium length, straight and well spread.

Female

With the exception of the tail, which is well spread and carried at an angle of 30° above the horizontal, the general characteristics are similar to those of the male, allowing for the natural sexual differences.

Colour

The black
Male and female plumage: The surface a lustrous green-black, and the undercolour slate or light grey.

In both sexes: Beak black, shading to yellow towards the tip. Eyes dark brown or hazel. Comb, face, wattles and ear-lobes red. Legs and feet black, with a tendency towards willow in adult birds, the underpart of the feet being yellow.

The white
Male and female plumage: The surface and undercolour white.

Jersey Giant

Jersey Giant large fowl
White male
Black female

Kraienköppe

The blue
Male and female plumage: Modern Langshan blue laced preferred.

In both sexes and all colours: Beak willow (some yellow permissible at present). Eyes dark brown to black. Comb, face, wattles and ear-lobes red. Legs and feet willow, i.e. dark greenish yellow, soles yellow. Skin nearly white.

Weights

Cock 5.90 kg (13 lb); cockerel 5.00 kg (11 lb)
Hen 4.55 kg (10 lb); pullet 3.60 kg (8 lb)

Scale of points

Shape and carriage	25
Colour	20
Quality	15
Head	10
Size and symmetry	10
Condition	10
Legs and feet	10
	100

Serious defects (in whites)

Smoky surface colour. Side sprigs to comb. More than 0.90 kg (2 lb) below standard weight in mature stock.

Defects (in black and whites)

Overhanging eyebrows. Sluggishness. Coarseness. Excessive or superfine bone. In blacks: black or dull black undercolour extending to the skin of the hackle, back, breast, or body and fluff. Positive white showing on surface of plumage. Other than yellow under the feet. More than 0.90 kg (2 lb) below the standard weight in mature stock.

Kraienköppe

Large fowl

Origin: Europe
Classification: Light
Egg colour: White

Kraienköppe (pronounced Cry-n-kerper) were originated on the German –Dutch border and were first shown in Germany in 1925. Bantams are more popular in this country than the large fowl.

Kraienköppe

Kraienköppe large fowl
Male

General characteristics: male

Carriage: Upright and elegant with powerful appearance.

Type: Body extended, strongly built, becoming fuller towards the rear. Back fairly long, straight, rounded at the sides with wide and abundant saddle hackle. Shoulders powerful and fairly wide. Breast wide and full. Wings long and powerful, carried closely with the tips under the saddle hackle. Tail fairly long and carried at an angle of 30/40°, with full sickles.

Head: Short, wide, arched, with a visibly prominent nape. Face free from feathers. Comb narrow walnut, in the shape of an elongated strawberry (or acorn), well set. Wattles short. Ear-lobes small. Beak short, strong, the tip bent downwards. Eyes fiery, alert, set somewhat under beetling brows.

Neck: Powerful, wide between the shoulders, of good average length, carried upright, curved slightly backwards with abundant hackle falling over the shoulders and back.

Legs and feet: Thighs powerful, prominent with smooth feathering. Shanks slender, smooth, free from feathers. Toes, four, fairly long, widely spread.

Plumage: Tight fitting.

Female

The general characteristics are similar to those of the male, allowing for the natural sexual differences. The back is carried almost horizontally. The tail is closed but not pointed. The comb is the size of a pea, very flattened. The wattles are small to the point of disappearance.

Kraienköppe

Colour

The silver
Male plumage: Head white, neck silvery white with black shaft stripe. Wing bows and back pure silvery white. Saddle silvery white with distinct shaft stripe. Wings: bays silvery white, wide black wing bars with green sheen, primaries black with narrow white outer lacing: secondaries' outer colour white, inner colour and tips black, so that the closed wing appears pure white. Breast, abdomen, thighs and hind part black, tail pure deep black with black sickles with green sheen.

Female plumage: Head silvery grey, hackle pure silvery white with black shaft stripe; back, shoulders and wings ash grey with silvery grey stippling and a whitish shaft. From hackle to tail every feather should show a narrow bright silvery grey lacing. Breast salmon to salmon red, abdomen and hind part ash grey: tail black and greyish black.

The golden
Male plumage: As the silver, except that the silvery white is replaced by golden red, lighter on head and neck.

Female plumage: Head and neck golden yellow marked as the silver. Back, shoulders and wings light brown ground colour of even shade with fine black striping, peppering and stippling; yellow shaft. From hackle to tail every feather should show a narrow golden lacing. Breast salmon to salmon red, abdomen and hind part brownish ash grey. Tail black with brown markings.

In both sexes and colours: Beak yellow with dark tip. Eyes yellow-red to red. Face, comb, ear-lobes and wattles red. Legs and feet bright yellow.

Weights

Male   2.50–2.95 kg (5½–6½ lb)
Female 1.80–2.50 kg (4–5½ lb)

Scale of points

Type	25
Colour	25
Head	20
Legs and feet	15
Condition	15
	100

Serious defects

Short or narrow body. Roach back. Upright or poorly-furnished tail. Low stance. Drooping wings. Thin neck. Coarse, pointed or narrow head. Fish eyes. Fluffy plumage. Narrow sickles. Any other comb.

Bantams

Kraienköppe bantams should follow the large fowl standard in every respect.

Weights

Male   850 g (30 oz)
Female 740 g (26 oz)

La Flèche

Large fowl

Origin: France
Classification: Heavy
Egg colour: Tinted

The La Flèche is a French breed which has never been widespread in Britain. A large black breed with two vertical spikes for a comb, it is related to the Crève-Cœur and in the middle of the nineteenth century was used to produce white-skinned *petit poussin* for the Paris market.

General characteristics: male

Carriage: Bold and upstanding.

Type: Body, general appearance large, powerful and rather hard. Back wide and rather long, slanting to the tail. Wings large and powerful. Breast full and prominent. Tail moderate in size.

Head: General appearance of the head long, slightly coarse and cruel. Beak large and strong with cavernous nostrils. Comb a double spike standing nearly upright with very small spikes in front. Wattles long and pendulous, ear-lobes large. Head should be quite free of crest.

Neck: Long and very upright, but not backward, with as much hackle as possible.

Legs and feet: Thighs and shanks long and powerful, the latter being free of feathers; toes large and straight.

Female

The general characteristics are similar to those of the male, allowing for the natural sexual differences.

Colour

Male and female plumage: Glossy black with bright green reflexions.

La Flèche

La Flèche large fowl Male

In both sexes: Beak black or very dark horn, comb, wattles and face deep red, ear-lobes brilliant white. Eyes bright red or black. Legs and feet very dark slate or leaden black.

Weights
Male   3.60–4.10 kg (8–9 lb)
Female 2.70–3.20 kg (6–7 lb)

Scale of points
Type and carriage	25
Head	35
Colour	15
Size	15
Legs and feet	5
Condition	5
	100

Serious defects
Presence of crest. Entirely red ear-lobes. Feathers on legs. Incorrect leg colour. Coloured feathers. Wry tail or any deformity.

Bantams

No bantam version is standardized.

Lakenvelder

Large fowl

Origin: Germany
Classification: Light
Egg colour: Tinted

As the name implies, this is a German breed. Though a useful breed utility-wise, its striking and handsome appearance has never really attracted the attention of English fanciers. The most striking feature of the bird is its plumage – black hackles and tail on a pure white body, and a good Lakenvelder plumage is as handsome as it is difficult to obtain.

General characteristics: male

Carriage: Upright, bold and sprightly.

Type: Body moderately long, fairly wide at the shoulders and narrowing slightly to the root of the tail. Full and round breast. Broad and apparently short back. Medium long wings, tucked well up, the bows and tips covered by the neck and saddle hackles. Long and full tail, the sickles carried at an angle of 45°, but avoiding 'squirrel' carriage.

Head: Skull short and fine. Beak strong and well curved. Eyes large, bright and prominent. Comb single, erect, evenly serrated, of medium size, and following the contour of the skull. Face smooth and of fine texture. Ear-lobes small and of almond shape. Wattles of medium length, well rounded at the base.

Neck: Of medium length and furnished with long hackle feathers flowing well on the shoulders.

Legs and feet: Of medium length. Thighs well apart. Shanks fine and round, free of feathers. Toes, four, strong and well spread.

Female

The general characteristics are similar to those of the male, allowing for the natural sexual differences. (*Note:* The comb is carried erect, and not drooping.)

Colour

Male and female plumage: Black and white. Neck hackle and tail solid black, free of stripes, ticks or spots. In the male the saddle hackle is white tipped with black. Remainder, including undercolour, pure white.

In both sexes: Beak dark horn. Eyes red or bright chestnut. Comb, face and wattles bright red. Ear-lobes white. Legs and feet slate blue.

Lakenvelder

Lakenvelder large fowl Male

Weights

Male   2.25–2.70 kg (5–6 lb)
Female 2.00 kg (4½ lb)

Scale of points

Colour	45
Size	20
Head	10
Type	10
Condition	10
Legs and feet	5
	100

Serious defects

Comb other than single. Feathers on shanks. Wry tail or any other deformity.

Bantams

No bantam version is standardized.

Leghorn

Large fowl

Origin: Mediterranean
Classification: Light
Egg colour: White

Italy was the original home of the Leghorn, but the first specimens of the white variety reached this country from America around 1870, and of the brown two years or so later. These early specimens weighed not more than 1.6 kg (3½ lb) each, but our breeders started to increase the body weight of the whites by crossing in the Minorca and Malay, until birds were produced well up to the weights of the heavy breeds. In the postwar years, the utility and commercial breeders established a type of their own, and that is the one which is now favoured. In commercial circles the white Leghorn has figured prominently in the establishment of high egg-producing hybrids.

General characteristics: male

Carriage: Very sprightly and alert, but without any suggestion of stiltiness or in-kneed appearance. Well balanced.

Type: Body wide at the shoulders and narrowing slightly to root of tail. Back long and flat, sloping slightly to the tail. Breast round, full and prominent, carried well forward; breast bone straight. Wings large, tightly carried and well tucked up. Tail moderately full and carried at an angle of 45° from the line of the back; full, sweeping sickles.

Head: Well balanced with fine skull. Beak short and stout, the point clear of the front of the comb. Eyes prominent. Comb single or rose. The single of fine texture, straight, and erect, moderately large but not overgrown, coarse or beefy, deeply and evenly serrated (the spikes broad at their base), extending well beyond the back of the head and following, without touching, the line of the head, free from 'thumb marks' and side spikes, or twist at the back. The rose moderately large, firm (not overgrown so as to obstruct the sight), the leader extending straight out behind and not following the line of the head, the top covered with small coral-like points of even height and free from hollows. Face smooth, fine in texture and free from wrinkles or folds. Ear-lobes well developed and pendent, equally matched in size and shape, smooth, open and free from folds. Wattles long, thin and fine in texture.

Neck: Long, profusely covered with hackle feathers and carried upright.

Legs and feet: Legs moderately long. Shanks fine and round – flat shins objectionable – and free of feathers. Ample width between legs. Toes, four, long, straight and well spread, the back toe straight out at rear. Scales small and close fitting.

Plumage: Of silky texture, free from woolliness or excessive feather.

Handling: Firm, with abundance of muscle.

Leghorn

Female

With the exception of the single comb rising from a firm base and falling gracefully over either side of face without obstructing the sight, and the tail, which is carried closely and not at such a high angle, the general characteristics are similar to those of the male, allowing for the natural sexual differences.

Colour

The black
Male and female plumage: Rich green-black or blue-black, the former preferred and perfectly free of any other colour.

The blue
Male and female plumage: Even medium shade of blue from head to tail, free from lacing, a dark tint allowed in the hackles of the male, but no black, 'sand' or any other colour than blue, and the more even the better.

The brown
Male and female plumage: Head and hackle rich orange-red striped with black, crimson red at the front of hackles below the wattles. Back, shoulder coverts and wing bow deep crimson red or maroon. Wing coverts steel blue with green reflexions forming a broad bar across; primaries brown; secondaries deep bay on outer web (all that appears when wing is closed) and black on the inner web. Saddle rich orange-red with or without a few black stripes. Breast and underparts glossy black, quite free from brown splashes. Tail black glossed with green; any white in tail is very objectionable. Tail coverts black edged with brown.
Female plumage: Hackle rich golden yellow, broadly striped with black. Breast salmon red, running into maroon around the head and wattles, and ash grey at the thighs. Body colour rich brown, very closely and evenly pencilled with black, the feathers free from light shafts, and the wings free from any red tinge. Tail black, outer feathers pencilled with brown.

The buff
Male and female plumage: Any shade of buff from lemon to dark, at the one extreme avoiding washiness and at the other a red tinge; the colour to be perfectly uniform, allowing for greater lustre on the hackle feathers and wing bow of the male.

The cuckoo
Male and female plumage: Light blue or grey ground, each feather barred across with bands of dark blue or grey, the markings to be uniform; the barring shading into the ground colour not cleanly cut but sharp enough to keep the two colours distinct.

The golden duckwing
Male plumage: Neck hackle rather light yellow or straw, a few shades deeper at the front below the wattles, the longer feathers striped with black. Back deep rich gold. Saddle and saddle hackle deep gold, shading in hackle to pale gold. Shoulder coverts bright gold or orange, solid colour (an admixture of lighter feathers is very objectionable). Wing bows the

Leghorn

Leghorn large fowl
Buff male
Brown female

Leghorn

**Leghorn large fowl
White male
Exchequer male**

Leghorn

same as the shoulder coverts; coverts metallic blue (blue-violet) forming an even bar across the wing, sharp, cleanly cut and not too broad; primaries black, with white edging on the outer web; secondaries white outer web (all that appears when the wing is closed), black inner and end of feather. Breast black with green lustre. Tail black, richly glossed with green-grey fluff at the base.

Female plumage: Head grey (a brown cap is very objectionable). Hackle white, each feather sharply striped with black or dark grey (a light tinge of yellow in the ground colour admitted). Breast and undercolour bright salmon red (this point is very important), darker on throat and shaded off to ash grey or fawn on the underparts. Back, wings, sides and saddle dark slate grey, finely pencilled with darker grey or black. Tail grey, slightly darker than the body colour, inside feathers dull black or dark grey.

The silver duckwing
Male plumage: Neck hackle silver white, the long feathers striped with black. Back, saddle and saddle hackle silver white. Shoulders and wing bow silver white, as solid as possible (any admixture of red or rusty feathers very objectionable). Wing coverts metallic blue (blue-violet) forming an even bar across the wing, which should be sharp and clearly cut, and not too broad; primaries black with white edging on outer parts; secondaries white outer edge (all that appears when the wing is closed), black inner and end of feathers. Thighs and underparts black. Tail black richly glossed with green, grey fluff at the base.

Female plumage: Head silver white. Hackle silver white, each feather sharply striped with black or dark grey. Breast and underparts light salmon or fawn, darker on throat and shaded off to ash grey on underparts. Back, wings, sides and saddle clear delicate silver grey or French grey, without any shade of red or brown, finely pencilled with dark grey or black (purity of colour very important). Tail grey, slightly darker than the body colour, with the inside feathers a dull black or dark grey.

The exchequer
Male and female plumage: Black and white evenly distributed with some white in the undercolour, the white of the surface colour in the form of a large blob as distinct from V-shaped ticking. Wings and tail to appear white and black evenly distributed.

The mottled
Male and female plumage: Black with white tips to each feather, the tips as evenly distributed as possible. Black to predominate and to have a rich green sheen.

The partridge
Male and female plumage: This colour is fully described under Wyandotte bantams and need not be repeated here.

The pile
Male plumage: Neck hackle bright orange. Back and saddle rich maroon. Shoulders and wing bows dark red. Secondaries dark chestnut outer web (all that appears when the wing is closed) and white inner. Remainder white.

Leghorn

Leghorn large fowl
Black male
Black female

Leghorn

Female plumage: Neck white tinged with gold. Breast deep salmon red shading into white thighs. Remainder white.

The white
Male and female plumage: Pure white free from straw tinge.

In both sexes and all colours: Beak yellow or horn. Eyes red. Comb, face and wattles bright red. Ear-lobes pure opaque white (resembling white kid) or cream, the former preferred. Legs and feet yellow or orange.

Weights

Cock 3.40 kg (7½ lb); cockerel 2.70–2.95 kg (6–6½ lb)
Hen 2.50 kg (5½ lb); pullet 2.00–2.25 kg (4½–5 lb)

Scale of points

Type	25
Comb	10
Lobe	10
Eyes	5
Legs	10
Breast	5
Size	5
Colour	20
Condition	10
	100

Serious defects in all varieties (for which a bird should be passed)

Single comb – Male's comb twisted or falling over, or female's erect. Ear-lobes red. Any white on face. Legs other than yellow or orange. Side sprigs on comb. Wry or squirrel tail, or any bodily deformity. *Rose comb* – Comb other than rose or such as to obstruct sight. Ear-lobes red. White in face. Wry or squirrel tail or any bodily deformity. Legs other than orange or yellow.

Bantams

Leghorn bantams should follow the large fowl standard in all respects.

Weights

Male 1020 kg (36 oz) max.
Female 910 kg (32 oz) max.

Malay

Large fowl

Origin: Asia
Classification: Heavy
Egg colour: Tinted

At the first poultry show in England in 1845 the Malay had its classification, and in the first British Book of Standards of 1865 descriptions were included of both the black-red and the white Malay. One of the oldest breeds, the Malay reached this country as early as 1830, and our breeders developed it, particularly in Cornwall and Devon. Today it is not widely bred, and is a purely exhibition breed.

General characteristics: male

Carriage: Fierce, gaunt, very erect, high in front, drooping at stern, straight at the hock, and a hard, clean, cut-up appearance from behind.

Type: Body wide fronted, short, and tapering; broad and square shoulders, the wing butts prominent, well up, and devoid of feathers at the point. Back short, sloping, and with convex outline, the saddle narrow and drooping. Breast deep and full, generally devoid of feathers at the point of the keel. Wings large, strong, carried high and closely to the sides. Tail of moderate length, drooping but not whipped, the sickles narrow and only slightly curved. The outline of the neck hackle, back, and tail (upper feathers) should form a succession of curves at nearly equal angles.

Head: Very broad with well-projecting or overhanging (beetle) eyebrows, giving a cruel and morose expression. Beak short, strong, and curved downwards. The profile of the skull and the beak approaches in shape a section of a circle. Eyes deep set. Comb shaped like a half-walnut, small, set well forward, and as free as possible from irregularities. Face smooth. Ear-lobes and wattles small.

Neck: Long and upright, with a slight curve, thick through from gullet to back of skull, the bare skin of the throat showing some way down the neck; the hackle full at the base of skull, but very short and scanty elsewhere.

Legs and feet: Legs long and massive, set well on the front of the body. Thighs muscular with very little feather, leaving the hocks perfectly exposed. Shanks free of feathers, beautifully scaled, flat at the hocks and gradually rounding to the spurs, which should have a downward curve. Toes, four, long, and straight, with powerful nails, the fourth (or hind) toe close to the ground.

Plumage: In all varieties is short and scanty, hard and narrow.

Handling: Firm fleshed and muscular.

Malay

Malay large fowl
Male
Female

Malay

Female

With the exception of the tail, which is carried slightly above the horizontal line, well 'played' as if flexible at the joint, rather short and square and neither fanned nor whipped, the general characteristics are similar to those of the male, allowing for the natural sexual differences.

Colour

The black
Male and female plumage: Glossy black all over with brilliant green and purple lustre, the green predominant, free from brassy or white feather.

The black-red
Male plumage: Hackle, saddle, back and wing bow rich red. Secondaries bright bay. Flights black inner web, red outside edging. Remainder lustrous green-black.
Female plumage: Any shade of cinnamon with dark purple-tinted hackle; quite free of ticks, spangles, or pencilling, or white in tail and wings. Partridge marked and clay females with golden hackle are also allowed.

The pile
Male plumage: Hackle, saddle, back, and wing bow rich red. Secondaries bright bay. Flights white inner web, red outside edging. Remainder cream white.
Female plumage: Hackle gold. Breast salmon. Remainder cream white.

The spangled
Male plumage: Breast, underparts, thighs and tail an admixture of red and white. Remainder, each feather somewhat resembling tortoise-shell in the blending of red or chestnut with black, and with a bold white tip or spangle, the flight feathers and tail as tri-coloured as possible.
Female plumage: Rich dark red or chestnut boldly marked with black and white.

The white
Male and female plumage: Pure white free from any yellow, black, or ruddy feathers.

In both sexes and all colours: Beak yellow or horn. Eyes pearl, white, yellow or daw with a green shade but the lighter the better; a red or foxy tinge very objectionable. Comb, face, throat, wattles and ear-lobes brilliant red. Legs and feet rich yellow, although in the black a slight duskiness may be overlooked.

Note: The foregoing are the principal varieties, and others are not kept or bred in sufficient numbers to warrant description. The above colours and markings are ideal, but type and quality are the most important points in the Malay fowl.

Weights

Male   5.00 kg (11 lb) approx.
Female 4.10 kg (9 lb) approx.

Scale of points

Type (shoulders 7, curves and carriage 16, reach 12)	35
Head	16
Eyes	9
Legs	10
Feathering	10
Colour	6
Tail	6
Condition	8
	100

Note: Size to be left to the discretion of the judge.

Serious defects

Any clear evidence of an alien cross. Lack of size, in large fowl, oversize in bantams. Single, spreading, or pea comb. Red eye, bow legs, knock knees, bad feet, in short, any defect not sufficiently penalized by the deduction of the maximum number of points allowed in the above scale.

Bantams

Malay Game is an old established bantam breed once popular in Cornwall and Devon. They are large in comparison with other bantams and it is not easy to reduce them further without losing the typical large breed characteristics. They should follow the large fowl standard in every respect.

Weights

Male 1190–1360 g (42–48 oz)
Female 1020–1130 g (36–40 oz)

Most specimens exceed these weights.

Marans

Large fowl

Origin: France
Classification: Heavy
Egg colour: Dark brown

Taking its name from the town of Marans in France, this breed has in its make-up such breeds as the Coucou de Malines, Croad Langshan, Rennes,

Marans

Faverolles, barred Rock, Braekel and Gatinaise. Imported into this country round about 1929, it has developed as a dual-purpose sitting breed. Like other barred breeds the cuckoo Marans females can be mated with males of other suitable unbarred breeds to give sex-linked offspring of the white head-spot distinguishing characteristic.

General characteristics: male

Carriage: Active, compact and graceful.

Type: Body of medium length with good width and depth throughout; front broad, full and deep. Breast long, well fleshed, of good width, and without keeliness. Tail well carried, high.

Head: Refined. Beak deep and of medium size. Eyes large and prominent; pupil large and defined. Comb single, medium size, straight, erect, with five to seven serrations, and of fine texture. Face smooth. Wattles of medium size and fine texture.

Neck: Of medium length and not too profusely feathered.

Legs and feet: Legs of medium length, wide apart, and good quality bone. Thighs well fleshed, but not heavy in bone. Shanks clean and unfeathered. Toes, four, well spread and straight.

Plumage: Fairly tight and of silky texture generally.

Handling: Firm, as befits a table breed. Flesh white, and skin of fine texture.

Female

General characteristics similar to those of the male, allowing for the natural sexual differences. Table and laying qualities to be taken carefully into account jointly.

Colour

The black
Male and female plumage: Black with a beetle green sheen.

The dark cuckoo
Male and female plumage: Cuckoo throughout, each feather barred across with bands of blue-black. A lighter shaded neck in both male and female, and also back in the male, is permissible if definitely barred. Cuckoo throughout is the ideal, as even as possible.

The golden cuckoo
Male plumage: Hackles bluish grey with golden and black bars; neck paler than saddle. Breast bluish grey with black bars, pale golden shading on upper part. Thighs and fluff light bluish grey with medium black barring. Back, shoulders and wing bows bluish grey with rich bright golden and black bars. Wing bars bluish grey with black bars: golden fringe permissible. Wings, primaries dark blue-grey, lightly barred; secondaries

Marans

Marans large fowl
Dark cuckoo male
Dark cuckoo female

dark blue-grey, lightly barred, with slight golden fringe. Tail dark blue-grey barred with black; coverts blue-grey barred with black. General cuckoo markings.
Female plumage: Hackle medium bluish grey with golden and black bars. Breast dark bluish grey with black bars, pale golden shading on upper parts. Remainder dark bluish grey with black bars. Cuckoo markings.

The silver cuckoo
Male plumage: Mainly white in neck and showing white on upper part of breast, also on top. Remainder barred throughout, with lighter ground colour than the dark cuckoo.
Female plumage: Mainly white in neck and showing white on upper part of breast. Remainder barred throughout, with lighter ground colour than the dark cuckoo.

In both sexes and all colours: Beak, white or horn. Eyes red or bright orange preferred. Comb, face, wattles and ear-lobes red. Legs and feet white.

Weights

Cock 3.60 kg (8 lb); cockerel 3.20 kg (7 lb)
Hen 3.20 kg (7 lb); pullet 2.70 kg (6 lb)

Scale of points

Type, carriage and table merits (to include type of breast and fleshing, also quality of flesh)	40
Size and quality	20
Colour and markings	15
Head	10
Condition	10
Legs and feet	5
	100

Serious defects

Feathered shanks. General coarseness. Lack of activity. Superfine bone. Any points against utility or reproductive values.

Defects (for which a bird may be passed)

Deformities, crooked breast bone, other than four toes, etc.

Defects (in blacks)

Restricted white in undercolour in both sexes. A little darkish pigmentation in white shanks.

Bantams

Marans bantams should be true miniatures of their large fowl counterparts. No black bantam variety is standardized.

Weights

Cock 910 g (32 oz); cockerel 790 g (28 oz)
Hen 790 g (28 oz); pullet 680 g (24 oz)

Marsh Daisy

Large fowl

Origin: Great Britain
Classification: Light
Egg colour: Tinted

This 'flowery' title was given to this breed when it was first created about 1910. Breeds in its make-up include Old English Game, Malay Game and Sicilian Buttercup. Its profitable laying powers stretch as far as the fourth season.

General characteristics: male

Carriage: Upright, bold and active.

Type: Body long, fairly broad, especially at the shoulders, with square and blocky appearance. Almost horizontal back. Well rounded and prominent breast. Full tail, carried at an angle of 45° from the vertical.

Head: Skull fine. Beak short and well curved. Eyes bold and prominent. Comb rose, medium size, well and evenly spiked, finishing in a single leader 1.25 cm (½ in) long in line with the surface, not as high as the Hamburgh's or following the nape of the neck as the Wyandotte's. Face smooth. Ear-lobes almond shaped. Wattles of fine texture and in keeping with the comb.

Neck: Fairly long, fine. Hackle flowing and falling well on the shoulders to form the cape.

Legs: Moderately long. Shanks and feet light boned, free from feathers. Toes, four, well spread.

Plumage: Semi-hard, of fine texture; profuse feathering to be deprecated.

Female

The general characteristics are similar to those of the male, allowing for the natural sexual differences.

Marsh Daisy

Colour

The black
Male and female plumage: Black, with beetle green sheen in abundance.

The buff
Male and female plumage: Golden buff throughout and buff to the skin. (*Note:* The male's tail is black to bronze but the ideal is a whole buff bird.)

The brown
Male plumage: Neck hackle rich gold, back and saddle dark gold. Main tail black; sickles black; coverts black, the whole to have a beetle green sheen. Saddle hackle dark gold, a little lighter gold at tips not objectionable. Wing bow dark gold, same shade as back; coverts or bar, black with beetle green sheen; secondaries forming the bay a flat brown, showing a triangular brown bay; primaries a flat black, with the lower edge flat brown, and all well hidden when the wing is closed and tucked up. Breast and all underbody parts black with patches of golden brown spangled in; solid shiny black should be striven for in these parts. Undercolour decided blue to blue-grey, with a little buff or light golden brown in places on breast.
Female plumage: Head and hackle rich gold, the tips of all feathers black, the whole to form a fringe at the cape. Back and wings brown ground ticked or peppered with darker brown or flat black; this may result in a series of black bars across the feathers, which is not objectionable. Tail dull flat black, a little lighter at the edge of the feathers not a disqualification, but should be discouraged. Breast and all underbody parts red wheaten or salmon, a level shade all over; too light a shade for these parts, or too deep a red wheaten, should not be striven for.

The wheaten
Male plumage: Hackles rich gold. Back and wing bow deep gold. Tail (coverts and sickles) rich beetle green-black. Remainder golden brown, the colour of a fairly dark bay horse. Undercolour (seen when the feathers are raised) from smoke white to a French or blue-grey, a little light buff fluff at the skin of the breast permissible.
Female plumage: Hackle chestnut with black tips forming a fringe at the base of it. Shoulders and back (upper part) red wheat; lower part of back to root of tail lighter shade, due to the feathers having a white wheat edging and red wheat centre and giving a dappling effect. Wing bows red wheat, the flights presenting a triangular patch of light brown when closed. Breast white wheat. Tail dull black with red wheat edging. Undercolour of back, smoke white to blue-grey; of breast, pure white.

The white
Male and female plumage: Pure white.

In both sexes and all colours: Beak horn. Eyes rich red with black pupil. Comb, face and wattles red. Ear-lobes white. Legs and feet pale willow green; toe-nails horn.

March Daisy

Marsh Daisy large fowl Male

Marsh Daisy large fowl Female

Minorca

Weights

Male 2.50–2.95 kg (5½–6½ lb)
Female 2.00–2.50 kg (4½–5½ lb)

Scale of points

Head (lobes 13, comb and wattles 10, other points 10)	33
Plumage	20
Condition	20
Type	15
Legs	12
	100

plus 'Laying power', 20

Serious defects

Want of type. Less than one-third white lobe. Red plumage, legs other than willow green.

Bantams

There is no bantam version of this breed standardized.

Minorca

Large fowl

Origin: Mediterranean
Classification: Light
Egg colour: White

The Minorca has been developed in this country as our heaviest light breed, and was at one time famous for its extra-large white eggs. Crossing with the Langshan and other heavy breeds did not improve the egg production of the breed, and concentration on exaggerated headgear had a similar effect. Those times are passed and wiser counsels now prevail. The result is that a much better balanced type is aimed for on the show-bench, with moderate size of lobes and of comb, and a more prominent frontal.

General characteristics: male

Carriage: Upright and graceful.

Minorca

Type: Body broad at the shoulders, square and compact. Breast full and rounded. Wings moderate in length, neat fitting close to body. Tail full, sickles long, well arched and carried well back.

Head: Long and broad, so as correctly to carry the comb quite erect. Beak fairly long and stout. Eyes full, bright and expressive. Comb single or rose in large fowl, single only in bantams. The single large, evenly serrated, perfectly upright, firmly set on the head, straight in front, free from any twist or thumb mark, reaching well to the back of the head, moderately rough in texture and free from any side sprigs. The rose oblong shape, broad at the base over the eyes, closely fitting, upright, firmly carried, full in front and tapering gradually to the 'leader' at the back, surface evenly covered with small nodules or points, free from hollowness, the 'leader' to follow the curve of the neck but not to touch the hackle. Face fine in quality, as free from feathers or hairs as possible, and not showing any white. Ear-lobes medium in size, almond shaped, smooth, flat, fitting close to the head. The lobe should not exceed 6.88 cm (2¾ in) deep and 3.75 cm (1½ in) at its widest part on the top, tapering as the Valencia almond in shape. No definite size for lobes is fixed in bantams. Wattles, long and rounded at the end.

Neck: Long, nicely arched, with flowing hackle.

Legs and feet: Legs of medium length, and thighs stout. Toes four.

Female

The general characteristics are similar to those of the male, allowing for the natural sexual differences, with the exception of the single comb which drops well down over the side of the face, so as not to obstruct the sight, and the ear-lobes which are 3.75 cm (1¾ in) deep and 3.13 cm (1¼ in) wide.

Colour

The black
Male and female plumage: Glossy black.

In both sexes: Beak dark horn. Eyes dark. Comb, face and wattles blood red. Ear-lobes pure white. Legs black or very dark slate, the latter in adults only.

The white
Male and female plumage: Glossy white.

In both sexes: Beak white. Eyes red. Comb, face and wattles blood red. Legs pinky white.

The blue
Male and female plumage: Soft medium blue, free from lacing, the male darker in hackles, wing bows and back.

In both sexes: Beak, comb, face, ear-lobes and wattles as in the black. Eyes dark brown (darker the better). Legs blue to slate.

Minorca

Minorca large fowl
Black male
Blue female

Minorca

Weights

Cock 3.20–3.60 kg (7–8 lb); cockerel 2.70–3.60 kg (6–8 lb)
Hen 2.70–3.60 kg (6–8 lb); pullet 2.70–3.20 kg (6–7 lb)

Scale of points

The black and white

Style, symmetry (type)	10
Size	15
Face	15
Comb	15
Ear-lobes	10
Legs, eyes and beak	8
Colour	10
Condition	10
Breast bone	7
	100

The blue

Type	15
Size	10
Condition	10
Head, comb and ear-lobes	5
Colour	60
	100

Serious defects

White or blue in face. Feathers on legs. Other than four toes. Wry or squirrel tail. Plumage other than black, blue or white in the several varieties. Legs other than black or slate in the black, blue or slate blue in the blue, or white in the white. Side sprigs in comb.

Bantams

Minorca bantams should be miniatures of their large fowl counterparts and standard points, colour and defects to be the same for bantams as for large fowl.

Weights

Male 960 g (34 oz)
Female 850 g (30 oz)
   Smaller specimens to be favoured, other points being equal.

Modern Game

Large fowl

Origin: Great Britain
Classification: Heavy
Egg colour: Tinted

By the introduction of Malay crosses, and with the skill of British fanciers, the Modern Game fowl was evolved. Black-reds, duckwings, brown-reds, piles, and birchens are the recognized varieties, the general characteristics being the same for each.

General characteristics: male

Carriage: Upstanding and active. In the show-pen the bird should show plenty of 'lift' as if reaching to its fullest height.

Type: Body short, flat back, wide front and tapering to the tail, shaped like a smoothing iron. Shoulders prominent and carried well up. Wings short and strong. Tail short, fine, closely whipped together, carried slightly above the level of the body, the sickles narrow, well pointed and only slightly curved.

Head: Long, snaky and narrow between the eyes. Beak long, gracefully curved and strong at the base. Eyes prominent. Comb single, small, upright, of fine texture, evenly serrated. Face smooth. Ear-lobes and wattles fine and small to match the comb.

(It is customary to dub Game males, to remove comb, ear-lobes, and wattles, and thus leave the head and lower jaw smooth and free from ridges.)

Neck: Long and slightly arched, fitted with 'wiry' feathers, but thin at the junction with the body.

Legs and feet: Legs long and well rounded. Thighs muscular. Shanks free of feathers. Toes, four, long, fine and straight, the fourth (or hind) toe straight out and flat on the ground, not downwards against the ball of the foot (or 'duck-footed'), which is most objectionable.

Plumage: Short and hard.

Female

The general characteristics are similar to those of the male, allowing for the natural sexual differences.

Colour

Colour is very important in Moderns. Varieties include black-reds, duckwings, piles, brown-reds, birchens, blacks, whites and blues. Self colours had been brought to great excellence a few years ago, but are now seldom seen.

Modern Game

Modern Game large fowl
Pile male
Brown-red female

143

Modern Game

Legs and beaks vary with the colour-varieties, from yellow in piles and whites through willow in black-reds to black in birchens. Eyes similarly vary from bright red to black. Combs and faces vary from bright red to dark purple and black. Leg colours are definitely 'tied' to each variety. Thus whites and piles must always have yellow legs, while shanks are willow in duckwings and black in brown-reds.

Colours are not so numerous as in Old English Game.

The birchen
Male plumage: Hackle, back, saddle, shoulder coverts and wing bows silver white, the neck hackle with narrow black striping. Remainder rich black, the breast having a narrow silver margin around each feather, giving it a regular laced appearance gradually diminishing to perfect black thighs.
Female plumage: Hackle similar to that of the male. Remainder rich black, the breast very delicately laced as in the male.

In both sexes: Beak dark horn. Eyes black. Comb, face, wattles and ear-lobes dark purple. Legs and feet black.

The black-red
Male plumage: Cap orange-red. Neck hackle light orange, free from black stripe. Back and saddle rich crimson. Wing bow orange; bar green-black; primaries black; secondaries rich bay on the outer edge, black on the inner and tips, the rich bay alone showing when the wing is closed. Remainder green-black.
Female plumage: Hackle gold, slightly striped with black, running to clear gold on the cap. Breast rich salmon, running to ash on thighs. Tail black, except the top feathers, which should match the body colour. Remainder light partridge-brown ground, very finely pencilled, and a slight golden tinge pervading the whole, which should be even throughout, free from any ruddiness whatever and with no trace of pencilling on the flight feathers.

In both sexes: Beak dark green. Eyes, comb, face, wattles and ear-lobes bright red. Legs willow.

The brown-red
Male plumage: Hackle, back and wing bow bright lemon, the neck hackle feathers striped down the centre with green-black, not brown. Remainder green-black, the breast feathers edged with pale lemon as low as the top of the thighs.
Female plumage: Neck hackle light lemon to the top of the head, the lower feathers being striped with green-black. Remainder green-black, the breast laced as in the male, the shoulders free from ticking and the back from lacing. (*Note:* There should be only two colours in brown-red Game, viz. lemon and black. In the male the lemon should be very rich and bright, and in the female light; the black in both sexes should have a bright green gloss known as beetle green.)

In both sexes: Beak very dark horn, black preferred. Eyes, comb, face, wattles, ear-lobes, legs and feet black.

The golden duckwing
Male plumage: Hackle cream white, free from striping. Back and saddle pale orange or rich yellow. Wings: bow pale orange or rich yellow; bars

and primaries black with blue sheen; secondaries pure white on the outer edge, black on inner and tips, the pure white alone showing when the wing is closed. Remainder black with blue sheen.

Female plumage: Hackle silver white, finely striped with black. Breast salmon, diminishing to ash grey on thighs. Tail black, except top feathers, which should match the body colour. Remainder French or steel grey, very lightly pencilled with black, and even throughout.

In both sexes: Beak dark horn. Eyes ruby red. Comb, face, wattles and ear-lobes red. Legs and feet willow.

The silver duckwing

Male plumage: Hackle, back, saddle, shoulder coverts and wing bows silver white. Secondaries pure white on the outer edge and black on the inner, with tips of bay, the white alone showing when the wing is closed. Remainder lustrous blue-black.

Female plumage: Hackle silver white, finely striped with black. Breast pale salmon, diminishing to pale ash grey on thighs. Tail black, except top feathers which should match the body colour. Remainder light French grey with almost invisible black pencilling.

In both sexes: Beak, etc. as in the golden variety.

The pile

Male plumage: Hackle one shade of bright orange-yellow. (Dark or washy hackles to be avoided.) Back and saddle rich maroon. Wing bow maroon; bar white and free from splashes; primaries white; secondaries dark chestnut on the outer edge and white on the inner and tips, the dark chestnut alone showing when wing is closed. Remainder pure white.

Female plumage: Hackle white, tinged with gold. Breast rich salmon red. Remainder pure white.

In both sexes: Beak yellow. Eyes bright cherry red. Comb, face, wattles and ear-lobes red. Legs and feet rich orange-yellow.

The self coloured: These are now seldom seen. Whites were standardized with red eyes and yellow legs. Blacks were permitted to have either red or dark purple faces and black legs. Blues had red faces, red eyes and blue-black legs.

Weights

Male   3.20–4.10 kg (7–9 lb)
Female 2.25–3.20 kg (5–7 lb)

Scale of points

Type and style	30
Head and neck	10
Eyes	10
Tail	10
Legs and feet	10
Colour	20
Condition and shortness of feather	10
	100

Modern Game

Modern game bantams
Black-red male
Pile female

Modern Game

Modern Game bantams
White male
Black-red female

Serious defects

Eyes other than standard colour. Flat shins. Crooked breast. Twisted toes or 'duck' feet. Wry tail. Crooked back.

Bantams

Modern Game bantams follow the standard for large fowl. Fine body, 'reachiness' and colour are the main points in bantams. The breed is the favourite of 'die-hard' showmen.

Weights

Male   570–620 g (20–22 oz)
Female 450–510 g (16–18 oz)

Modern Langshan

Large fowl

Origin: Asia
Classification: Heavy
Egg colour: Brown

The Modern Langshan has been developed on different lines from the Croad, and is quite another type of bird. It is much longer in the shanks and of a totally different outline from the original Croad.

General characteristics: male

Carriage: Graceful, upright, alert, strong on the leg with the bearing of an active bird.

Type: Body long and broad but by no means deep. Back horizontal when in normal attitude, and with close compact plumage. Shoulders broad and abundantly furnished saddle. Wings large, closely carried, but neither 'clipped' nor 'pinched in'. Tail full, flowing, spread at base, carried fairly high but not squirrel, furnished with abundant side hangers and two sickles, each feather tapering to a point.

Head: Fine. Beak fairly long and slightly curved. Eyes large. Comb single, straight, upright, fairly small, and evenly serrated with five or six spikes. Face of fine texture. Ear-lobes medium size, pendent and inclined to fold. Wattles medium length, fine texture, neatly rounded.

Neck: Fairly long, broad at base and covered with full hackle.

Legs and feet: Legs rather long, strong and wide apart. Thighs covered with closely fitting feathers, especially around the hocks. Shanks strong but not coarse boned, with an even fringe of feathers (not heavy) on the outer sides. Toes, four, long, straight and well spread, the outer toe (and that alone) slightly feathered.

Plumage: Close and smooth.

Female

With the exception of the tail (not carried high) the general characteristics are similar to those of the male, allowing for the natural sexual differences.

Colour

The black
Male and female plumage: Black with a brilliant beetle green sheen.

In both sexes: Beak dark horn to black. Eyes dark brown to black, the darker the better. Comb, face, wattles and ear-lobes brilliant red. Legs and feet dark grey with black scales in front and down the toes, showing pink between the scales (especially down the outer sides of the shanks) and on the skin between the toes. Toe-nails white. Underfoot pink-white. Skin of the body and thighs white and transparent.

The blue
Male plumage: Hackles, back, tail, sickles, side hangers, and wing bow rich deep slate, the darker the better, with brilliant purple sheen. Remainder clear slate blue, each feather distinctly laced (edged) with the same dark shade as the back, the contrast between delicate ground and dark lacing being well defined.

Female plumage: Head and upper part of neck rich dark slate. Remainder clear slate blue, each feather distinctly laced (edged) with dark slate, the lacing being well defined.

In both sexes: Beak, etc. as in the black.

The white
Male and female plumage: Pure white, with brilliant silver gloss.

In both sexes: Beak white with a pink shade near the lower edges. Legs and feet light grey or slate showing pink between the scales and on the skin between the toes. Eyes, comb, face, wattles, ear-lobes, toe-nails, underfoot and skin as in the black.

Weights

Cock 4.55 kg (10 lb); cockerel 3.60 kg (8 lb)
Hen 3.60 kg (8 lb); pullet 2.70 kg (6 lb)

Scale of points

Type and carriage	35
Head	15
Legs and feet	10
Size	10
Colour and plumage	20
Condition	10
	100

Modern Langsham large fowl
Female

Serious defects

Yellow skin. Yellow base of beak. Yellow or orange-coloured eyes. Yellow around the eyes. Yellow shanks or underfoot. Legs other than standard colour. Shanks not feathered. More than four toes. Permanent white in the face or ear-lobes. Comb with side sprigs, or other than single. Wry or squirrel tail. Coloured feathers.

Defects

Absence of pink between toes. Feathering on middle toes. Outer toe not feathered. Too scantily or too heavily feathered shanks or outer toes. Twisted toes. Short shanks. Crooked breast. Twisted or falling comb. General coarseness. Too much fluff. Purple sheen in black; yellow shade in white.

Bantams

Modern Langshans are not seen in bantam form. Should they appear then they should follow the large fowl standard.

Nankin Bantams

Origin: Asia
Classification: True bantam
Egg colour: Tinted

General characteristics: male

Carriage: Jaunty and active, with a proud bearing.

Type: Body small and neat. The breast is carried well up and forward. Back, short and sloping to the tail. Wings large, closely folded and carried very low, almost down to the ground. Tail large and well spread, long, well-curved sickles sweeping right round in a circle. The whole tail carried very high, but not squirrel.

Head: Small and fine. Beak longish, rather fine, and only slightly curved. Eyes very large and bright. Comb single or rose. The single should be small and neat, bearing three to six fine spikes. The comb should be straight and upright, carried well away from the head, a flyaway comb being characteristic. The rose comb should be small and close fitting, finely worked, with a small leader curved gracefully upward. Face smooth, ear-lobes very small. Wattles small and well rounded, fine and smooth.

Neck: Medium long, well curved and bearing long and abundant hackle.

Legs and feet: Legs short, thighs set well apart. Shanks rather short rounded and fine, free from feathers. Toes, four, rather small, straight and well spread.

Female

With the exception of the carriage, which is lower and less arrogant, with the tail carried well spread but lower, the general characteristics are similar to those of the male allowing for the natural sexual differences.

Colour

Male plumage: Back, wing coverts, neck and saddle hackles a rich orange or chestnut. Tail, the main feather black, sickles and side hangers chestnut or copper, shading into black. Remainder of the plumage ginger buff.

Female plumage: Neck, back wings and saddle dark ginger buff. Tail brown or dark ginger buff, shading into black at the ends. Remainder of the plumage light, ginger buff.

In both sexes: Beak white. Eyes bright orange. Comb, face, lobes and wattles bright cherry red. Legs blue or bluish white.

Weights

Male   740–850 g (26–30 oz)
Female 680–790 g (24–28 oz)

New Hampshire Red

Scale of points

Type and carriage	30
Colour and markings	30
Feet and legs	10
Comb	10
Other head points	10
Condition	10
	100

Serious defects

Wry or squirrel tail. Comb other than rose or single. Comb, lopped or twisted or curved down at the rear. Long legs, large feet, or twisted toes. White in the lobes or face. Yellow legs. Even colour throughout. White in the base or the tail. All black or buff tails in males, tail lacking black ends in females. Visible black in the closed wing. Any physical deformity.

Minor faults

Black in the inner web of the wing; blue beak.

New Hampshire Red

Origin: America
Classification: Heavy
Egg colour: Tinted to brown

As if to copy the farmers in the State of Rhode Island who developed the breed carrying its name, those in the neighbouring State of New Hampshire developed and named their breed. The New Hampshire Red was bred by selection from the Rhode Island Red without the introduction of any other breed, taking some thirty years to reach standardization in 1935. Early maturity, quick feathering, and a plump carcase are particular features of the breed. Its body shape and colouring are very different from those of the Rhode Island Red.

General characteristics: male

Carriage: Active and well balanced.

Type: Body of medium length, relatively broad, deep and well rounded. Back of medium length, broad for entire length, gradual concave sweep to tail. Breast deep, full, broad and well rounded, the keel relatively long and extending well to the front at the breast. Wings moderately large, well folded, carried horizontally and close to the body, fronts well covered by breast feathers; primaries and secondaries broad and overlapping in natural order when wing is folded. Tail of medium length, well spread and

New Hampshire Red

New Hampshire Red
large fowl
Male
Female

New Hampshire Red

carried at angle of 45°. Sickles medium in length, extending well beyond the main tail; lesser sickles and coverts medium length and broad; main tail feathers broad and overlapping.

Head: Of medium length, fairly deep, inclined to be flat on top rather than round. Beak strong, medium length, regularly curved. Eyes large, full and prominent, moderately high in the head. Comb single, medium size, well developed, set firmly on head, perfectly straight and upright, having five well-defined points, those in front and rear smaller than those in centre; the blade smooth and inclining slightly downward, not following too closely the shape of the neck. Face smooth, full in front of the eyes; skin of fine texture. Wattles moderately large, uniform, free from folds or wrinkles. Ear-lobes elongated oval, smooth and close to head.

Neck: Of medium length, well arched, hackle abundant, flowing well over shoulders. Moderately close feathered.

Legs and feet: Legs well apart, straight when viewed from the front. Lower thighs large, muscular and of medium length. Toes, four, medium length, straight and well spread.

Plumage: Feather character to be of broad firm structure, overlapping well and fitting tightly to the body. Fluff moderately full.

Female

The general characteristics are similar to those of the male, allowing for the natural sexual differences. Comb slightly tilted at rear. Wattles of medium size, well developed and well rounded. Tail moderately well spread, carried at angle of 35°. Wings rather large carried nearly horizontal.

Colour

Male plumage: Head brilliant reddish bay. Breast and neck medium chestnut red. Back brilliant deep chestnut red. Saddle rich brilliant reddish bay, slightly darker than neck. Wing fronts medium chestnut red, bows brilliant chestnut red; coverts deep chestnut red; primaries, upper web medium red, lower web black edged with medium red; primary coverts black edged with medium red; secondaries, upper web medium chestnut red having broad stripe extending along shaft to within inch of tip, lower web medium chestnut red; shaft red. Tail, main feathers black; sickles rich lustrous greenish black; coverts lustrous greenish black edged with deep chestnut red; lesser coverts deep chestnut red. Body and fluff medium chestnut red. Lower thighs medium chestnut red. Undercolour in all sections light salmon, a slight smoky tinge not a defect.

Female plumage: Head medium chestnut red. Neck medium chestnut red, each feather edged with brilliant chestnut red; lower neck feathers distinctly tipped with black; feathers in front of neck medium chestnut red. Wing fronts, bows and coverts medium chestnut red; primaries, upper web medium red, lower web medium red with narrow stripe of black extending along shaft; shaft medium red; primary coverts black edged with medium red; secondaries, lower web medium chestnut red, upper web medium chestnut red with black marking extending along edge of shaft two-thirds its length. Back, breast, lower thighs, body and fluff medium chestnut red. Tail, main feathers black edged with medium chestnut red; shaft medium chestnut red. Undercolour as in male.

In both sexes: Beak reddish brown. Eyes bay. Comb, face, wattles and ear-lobes bright red. Legs and toes rich yellow, tinged with reddish horn. Line of reddish pigment down sides of shanks extending to tips of toes desirable in male.

Weights

Cock 3.85 kg (8½ lb); cockerel 3.40 kg (7½ lb)
Hen 2.95 kg (6½ lb); pullet 2.50 kg (5½ lb)

Scale of points

Type and carriage	25
Colour	20
Dual-purpose quality	15
Head	10
Size and symmetry	10
Legs and feet	10
Condition	10
	100

Norfolk Grey

Origin: Great Britain
Classification: Heavy
Egg colour: Tinted

The Norfolk Grey was first introduced by Mr. Myhill of Norwich under the ugly name of Black Marias. They were first shown at the 1920 Dairy Show and were mainly the result of a cross between silver birchen Game and duckwing Leghorns. They now occasionally appear at shows and are to be seen around their county of origin.

General characteristics: male

Carriage: Fairly upright and very active.

Type: Body rather long; broad at the shoulders. Full round breast carried upwards. Large wings well tucked up. Well-feathered tail.

Head: Skull fine. Beak short and well curved. Eyes large and bold. Comb single, upright, of medium size, well serrated, and with a firm base. Face smooth and fine. Ear-lobes small and oval. Wattles long and fine.

Neck: Of medium length, abundantly covered with hackle.

Norfolk Grey

Norfolk Grey large fowl Male

Legs and feet: Fairly short and set well back. Shanks free from feathers. Toes, four, well spread.

Plumage: Close.

Female

The general characteristics are similar to those of the male, allowing for the natural sexual differences.

Colour

Male plumage: Neck, back, saddle, shoulder coverts and wing bars silver white, the hackles with black striping, free from smuttiness. Remainder a solid black.

Female plumage: Hackle similar to that of the male. Remainder black, the throat very delicately laced with silver (about 5 cm (2 in) only).

In both sexes: Beak horn. Eyes dark. Comb, face, ear-lobes and wattles red. Legs and feet black or slate black, the former preferred.

Weights

Male   3.20–3.60 kg (7–8 lb)
Female 2.25–2.70 kg (5–6 lb)

North Holland Blue

Scale of points

	Male	Female
Colour and markings: hackles	20	15
Colour and markings: back and wings	10	15
Colour and markings: breast, thighs and fluff	15	15
Type and size	20	20
Head (comb and lobes 10, eyes 5)	15	15
Legs and feet	5	5
Condition	10	10
Tail	5	5
	100	100

Serious defects

White in lobes. Comb other than single or obstructing the sight. Legs other than black or slate black. Feathers on shanks or feet. Lacing or shaftiness on back, breast, or wings of females.

North Holland Blue

Origin: Holland
Classification: Heavy
Egg colour: Tinted

The blood of the Malines in the make-up of this breed is seen in its quick maturity and rapid growth. To further its commercial importance its utility properties are valued on the show-bench in preference to markings. Lightly feathered shanks are a standardized characteristic, and the male is lighter in colour than the female. As a barred breed, the females when mated with unbarred males of breeds with dark downs, produce sex-linked offspring, the male chicks having the white head-spot when hatched, which is absent in the female chicks.

General characteristics: male

Carriage: Upright, bold and alert.

Type: Body substantial in build, yet compact. Back broad, flat, horizontal, reasonably long, with slight rise to tail, broad saddle, prominent shoulders. Breast full, rounded and prominent; breast bone long, well

North Holland Blue

fleshed and rounded off, neither shallow nor keely. Well-rounded sides, good depth of body, with a well-developed abdomen. Wings strong, well developed and close to the body. Tail broad, short, well spread with medium furnishings.

Head: Rounded and of medium length. Beak stout and short. Eyes full, bold, with keen expression, the pupils well formed and large. Face smooth, full, fine texture and without heavy eyebrows. Comb single, upright, of medium size and fine texture, with five to seven neat even serrations, following slightly the curve of the neck at the back. Ear-lobes of medium size and silky. Wattles medium in length and size, and fine in texture.

Neck: Somewhat broad, medium in length and not too profusely feathered.

Legs and feet: Legs of medium length, wide apart, well formed and good quality bone with close scales. Thighs well developed and fleshed. Toes, four, straight, and well spread. Shanks lightly feathered, including outer toe.

Plumage: In general not too profuse. Fine in texture.

Handling: Firm, as befits a table breed. Skin thin and of fine texture. Flesh high-quality table grade.

Female

With the exception of the shanks, which are more heavily feathered, the general characteristics are similar to those of the male, allowing for natural sexual differences.

Colour

Male plumage: Lighter blue-grey than female, barred. Undercolour very pale blue-grey, barring immaterial.

Female plumage: Dark grey-blue, slightly barred. Undercolour paler blue-grey, barring immaterial.

In both sexes: Beak white superimposed blue. Eyes orange to red. Comb, wattles, face and ear-lobes red. Shanks white, but may be shaded with blue. Skin white.

Weights

Cock 3.85–4.80 kg (8½–10½ lb); cockerel 3.40–4.30 kg (7½–9½ lb)
Hen 3.20–4.10 kg (7–9 lb); pullet 2.70–3.60 kg (6–8 lb)

Scale of points

Type and carriage	10
Indication of table merits	20
Indication of egg production merits	20
Colour and markings	20

(continued)

North Holland Blue

North Holland Blue
large fowl
Male
Female

(**Scale of points** – *continued*)

Head	10
Legs and feet	10
Condition	10
	100

Defects

General coarseness. Superfine bone. Unfeathered shanks. Any points against table, laying or reproductive qualities. Birds may be passed for deformities, crooked breast bone, serious defects, and clean shanks devoid of any feathering.

Old Dutch Bantams

Origin: Holland
Classification: True bantam
Egg colour: Tinted

The Old Dutch bantam first appeared in this country about 1970. It was slow to gain supporters due to inbreeding but now is becoming more popular. It is possibly the smallest bantam around and found in several colours.

General characteristics: male

Carriage: Upright.

Type: Back very short, broad at shoulder, slightly sloping, saddle short and broad with abundant hackle running smoothly into tail coverts. Breast carried high, full and well forward. Wings relatively large and long but not too pointed, carried low and close. Tail upright, full and well spread, with well developed and curved sickles.

Head: Small, face smooth. Comb single, very small with five serrations tending towards flyaway type. Beak very short and strong, slightly curved. Eyes very large and lively. Wattles fine, short and round. Ear-lobes very small and fine, oval to almond shape.

Neck: Very short, curved, and finely tapered with plentiful hackle.

Legs and feet: Legs well spaced and straight, thighs short, shanks very short and free of feather. Toes, four, well spread.

Plumage: Luxuriant and lying close to body with plentiful sickles, side hangers and coverts.

*Old Dutch bantam
Gold duckwing male*

Female

The general characteristics are similar to those of the male, allowing for the natural sexual differences.

Colour

The partridge

Male plumage: Head orange reddish brown. Neck hackle a gradual transition from orange to light orange yellow, each feather having a greenish black middle stripe. Back deep reddish brown; side hangers corresponding with the neck hackle, a little darker permitted. Breast black with green sheen, free from markings or spots. Wing bow black, shoulders deep reddish brown, wing band iridescent greenish black; inner web black; outer web chestnut brown; internal web and tip black, tip being brown when the wing is closed. Flanks and thighs deep black with green sheen, free of markings and spots. Underpart and rear black. Tail, main feathers, sickles and tail coverts green iridescent black; the tail coverts nearest the side hangers with a brownish edge underneath at the tip. Undercolour greyish.

Female plumage: Head gold brown. Throat greyish white. Neck hackle goldish yellow, with a black middle stripe. Wings, back, saddle and tail coverts greyish brown with fine black pencillings, as even as possible, free of rust or red. Tail, main feathers blackish, the top feather on each side with brown pencillings. Breast light salmon brown, shading to brownish grey near the thighs. Thighs, flanks and down ash grey.

The silver duckwing
Male plumage: Exactly the same as the partridge coloured male in the black feathered parts and in the markings on the neck and saddle hangers. The orange and light orange-yellow and the brown are completely replaced by silvery white.
Female plumage: Head silver white. Hackle silver white, each feather having a black middle stripe. Wing coverts, back, saddle and tail coverts muted silver or slate grey with fine black pencillings as even as possible, free of flecks, rust brown or yellow. Tail feathers blackish, the top feather on each side silver pencilled. Breast light salmon-coloured brown fading to ash grey underneath. Rump and rear, underpart and flanks ash grey. Thighs ash grey with some pencilling.

The gold duckwing
Male plumage: Exactly the same as the silver duckwing male, except head rich straw yellow. Hackle and side hangers, neck light straw yellow, saddle a little darker. Back and shoulders deep orange-red.
Female plumage: Exactly the same as the silver duckwing female in all parts. They should, however, have a slightly warmer and mainly darker shade in the back, wing and saddle.

The blue duckwing
Male plumage: Exactly the same as the partridge coloured male with the exception of the black feathers and feather sections which, for this colour type, should be lighter and as evenly blue as possible. The down should be lighter blue than in the partridge coloured male.
Female plumage: Exactly the same as the partridge coloured female. The black feathers and feather sections must be lighter blue; the pencilling is blue.

The salmon
Male plumage: The male is exactly the same in colour and markings as the gold duckwing apart from a somewhat more irregular colour on the shoulders and back.
Female plumage: Head silver white or cream. Hackle ivory, white or cream, each feather having a black middle stripe; feathers on the front, creamy wheat coloured. Wing bow and wing coverts light cream wheat; outer web wheat coloured, inner web wheat grey with brownish black pencillings. Back and saddle light cream wheat coloured. Breast cream wheat coloured. Thighs, rump and rear, light cream wheat coloured. Tail, the uppermost feather on each side darker cream coloured with black spots, upper feathers dull black either with or without a cream-coloured edge along the underside and the tip.

In both sexes and all colours: Beak dark horn or bluish. Eyes orange-red to brownish red. Comb, face and wattles red. Ear-lobes pure white. Legs and feet slate blue.

Weights

Male   510–680 g (18–24 oz)
Female 400–570 g (14–20 oz)

Scale of points

Type and carriage	20
Feathering	15
Colour and markings	25
Head points	20
Legs and feet	10
Condition	10
	100

Serious defects

Long narrow build. Carriage too sloping. Large glossy white or reddish lobes. White in face. Wrong coloured legs or eyes. Whipped tail. Undeveloped ornamental feathers in male. Any splashing or spotting in plumage.

Old English Game

Origin: Great Britain
Classification: Light
Egg colour: Tinted

When the Romans invaded Britain, Julius Caesar wrote in his commentaries that the Britons kept fowls for pleasure and diversion but not for table purposes. Many well-known authorities have considered that cock fighting was the diversion. In 1849 an Act of Parliament was passed making cock fighting illegal in this country, and with poultry exhibitions then taking root, many breeders began to exhibit Game fowls. Over thirty colours of Old English Game have been known.

General characteristics: male

Carriage: Bold, sprightly, the movement quick and graceful, as if ready for any emergency.

Type: Back short and flat, broad at the shoulders, and tapering to the tail. Breast broad, full, and prominent with large pectoral muscles, and breast bone not deep or pointed. Wings large, long and powerful with large strong quills, amply protecting the thighs. Belly small and tight. Tail large, carried upwards and spread, main feathers and quills large and strong.

Head: (It is customary to dub O.E.G. cockerels.) Small and tapering. Beak big, boxing (i.e. the upper mandible shutting tightly and closely over the lower one), crooked or hawk-like, pointed, strong at the base. Eyes large, bold and full of expression. Comb single, small, thin, upright, of fine texture in undubbed males and females. Face and throat skin flexible and loose. Ear-lobes and wattles fine, small and thin.

Neck: Large boned, round, of fair length, and very strong at the junction with the body, furnished with long and wiry feathers covering the shoulders.

Legs and feet: Legs strong. Thighs short, round and muscular, following the line of the body, or slightly curved. Shanks strong, clean boned, sinewy, close scaled (not flat and 'gummy'), not stiffly upright or too wide apart, and with a good bend or angle at the hock, fitted with hard and fine spurs set low. Toes, four, thin, straight and tapering, terminating in long, strong, curved nails, the fourth (or hind) toe strong, straight out and flat on the ground.

Plumage: Hard and glossy, without much fluff.

Handling: Well balanced, hard yet light fleshed, 'corky', with plenty of muscle, and strong contraction of the wings and thighs to the body.

Female

With the exception of the tail, which is inclined to fan-shape and carried well up, the general characteristics are similar to those of the male, allowing for the natural sexual differences.

Colour

The following colours are recognized by the Oxford O.E.G. Fowl Club:

The black-breasted black-red
Male plumage: Hackle, shoulders and saddle rich dark red (the colour of the shoulders of a black-breasted red). The rest of the plumage black.
Female plumage: Body brown, mixed with umber brown, hackle striped red, breast red-brown, tail and primary wing feathers black.
In both sexes: Fluff (i.e. the down at the roots of the feathers next to the skin) black. Eyes, beak, legs and nails black. Face gypsy or purple.

The black-breasted red
Male plumage: Breast, thighs, belly and tail black; wing bars steel blue, secondaries (when closed) bay; hackle and saddle feathers orange-red; shoulders deep crimson-scarlet.
Female plumage: Hackle golden, lightly striped with black; breast robin; belly ash grey; back, shoulders and wings a good even partridge, primaries dark, also tail.
In both sexes: The dark legged birds should have grey fluff, the white and yellow legged, white fluff. Face scarlet red. Legs willow, yellow white carp or olive.

The shady or streaky-breasted light red
Male plumage: Hackle and back a shade lighter than the black-breasted red male and sometimes red wing bars.

Female plumage: Wheaten, a pale cream colour (like wheat) with clear red hackle; tail and primaries nearly black. The red wheaten (the colour of red wheat), or light brick red in body and wings; hackle dark red; tail dark.

In both sexes: Fluff white. Legs white or yellow.

The black-breasted silver duckwing
Male plumage: Resembles the black-breasted red in his black markings and blue wing bars; rest of the plumage clear silvery white.
Female plumage: Hackle white, lightly striped black; body and wings even silvery grey; breast pale salmon; primaries and tail nearly black.

In both sexes: Fluff light grey. Face red, eyes pearl. Legs and beak white. Or eyes red and legs dark.

The black-breasted yellow duckwing
Male plumage: Hackle and saddle yellow straw; shoulders deep golden; wing bars steel blue; secondaries white when closed; rest of plumage black.
Female plumage: Breast deeper, richer colour and body slightly browner tinge than the silver female.

In both sexes: Fluff light grey. Face red. Legs yellow, willow or dark.

The black-breasted birchen duckwing
Male plumage: Hackle deep rich straw, may be lightly striped; shoulders maroon; otherwise same as preceding.
Female plumage: Shade darker than yellow duckwing female; hackle more heavily striped with black, and often foxy on the shoulders.

In both sexes: Face slightly darker than in yellow duckwing. Legs yellow or dark.

The black-breasted dark grey
Male plumage: Like the black-breasted red, except hackle, saddle and shoulders a dark silver grey, often striped with black.
Female plumage: Nearly black, with grey striped hackle, or body very dark grey.

In both sexes: Fluff black. Beak, eyes and legs black. Face gypsy or purple.

Other greys may have laced, streaked or mottled grey or throstle breasts; hackle, saddle and shoulders more or less striped with black; legs and eyes dark; the females dark grey to match; fluff in both sexes, light or dark grey.
Note: Greys all differ from duckwings in having the secondaries, when closed, black; or, if grey, wanting the steel blue bar across them.

The clear mealy-breasted mealy grey
Male plumage: Nearly white breasted, with hackle and saddle the same, lightly striped; plumage and most of the tail grey.
Female plumage: Light grey.

In both sexes: Fluff light grey. Eyes and legs dark.

The brown-breasted brown-red
Male plumage: Breast, thighs, belly and closed wing mahogany brown; hackle and saddle almost similar; shoulders crimson; primaries and tail black or dark bronze brown.

Old English Game

Female plumage: Dark mottled brown with light shafts to the feathers.

In both sexes: Fluff black. Face deep crimson or purple. Eyes and legs dark.

The streaky-breasted orange-red
Male plumage: Breast streaked, laced or pheasant, black, marked with brown or copper colour; hackle and saddle brassy or coppery orange colour; shoulders crimson; the rest of the wings and the tail black.

Female plumage: Black or nearly black body, with tinsel hackle striped with black, or dark mottled brown and gold striped hackle.

In both sexes: Fluff black or nearly so. Face, eyes and legs dark.

The ginger-breasted ginger red
Male plumage: Breast and thighs deep yellow ochre, either clear or slightly pencilled or spotted; hackle and saddle red golden; shoulders crimson red; tail and flight feathers bronzy.

Female plumage: Golden yellow throughout, pencilled or spangled, particularly on back and wings, with bronze; tail pencilled bronze or dark.

In both sexes: Fluff dark. Beak, legs and eyes dark or yellow. Face purple or crimson.

The dun-breasted blue dun
Male plumage: Breast, belly, thighs, tail and closed secondaries the colour of a new slate, sometimes the breast marked with the same colour two shades darker; hackle, saddle and shoulders, and sometimes the tail coverts and the primaries two shades darker (like a slate colour after being wetted).

Female plumage: Blue slate colour, with dark hackle like the male, often marked or laced all over with the darker shade.

In both sexes: Fluff slate blue. Eyes, face and legs dark.

The streaky-breasted red dun
Male plumage: Breast slaty, streaked with copper red; hackle and saddle striped with slate or dark striped; shoulders crimson; wing bars and closed secondaries slate, or marked a little with brown; tail slaty or dark blue.

Female plumage: Body slaty all over, or laced in a darker shade; hackle golden striped, and sometimes marked with gold on the breast.

In both sexes: Fluff dark slate. Legs dark or yellow.

The yellow, silver and honey dun
These are coloured respectively with the following colours; the colour of new honeycomb is intended to describe the honey dun. They may have yellow or dark legs according to body colour, and white legs are permissible in the silver dun, as well as other coloured legs. The females are blue bodied with hackles to match their males. Smoky duns are of a dull smoke colour throughout; legs and eyes should be dark.

The pile
Male plumage: The smock-breasted blood wing pile is marked exactly like the black-breasted light red, except that the black and the blue wing bars are exchanged for a clear cream white. The breast may be streaked with red in red pile.

Old English Game

Old English Game
large fowl
Brown-red male
Blue dun female

Old English Game

Old English Game
large fowl
Tassel pile male
Grey female

Female plumage: White, with salmon breast and golden striped hackle, or streaked all over lightly with red.

In both sexes: Face and eyes red. Legs white, yellow or willow.

Note: Other varieties of piles may be streaky, marbled or robin breasted; and light lemon or custard in top colour, or dun piles having slate blue markings in place of red. All piles have white fluff.

The spangled
Male and female plumage: These have white tips to their feathers. The more of these spots and the more regularly they are distributed the better. The male should show white ends to the feathers on hackle and saddle. The ground colour may be red, black or brown, or a mixture of all three. Underfluff white.

In both sexes: Eyes and face red; legs any colour or mottled to match plumage.

The white
Male and female plumage: This variety should be free from any coloured feathers. Fluff pure white.

In both sexes: Beak and legs white. Face red; eyes pearl; or yellow legs and red eyes.

The black
Male and female plumage: This variety should be free from any white or coloured feathers and should possess black fluff.

In both sexes: Dark beaks, faces and legs and black eyes, though red faces and red eyes are allowed at present.

The furnace, brassy back and polecat
Male and female plumage: These are blacks with brass colour on their wings or back, and occasionally have yellow legs, which are allowed. The females are chiefly black, but often much streaked with grey-brown on breast and wings. Polecats are streaked with dark tan colour on hackles and saddle in the males. Legs dark.

The cuckoo
Male and female plumage: Cuckoo-breasted cuckoo resembles the Plymouth Rock fowl in markings of a blue-grey barred plumage.

In both sexes: Faces and eyes red; legs various.

Variations of this colour are yellow cuckoos, also creles, creoles, cirches, mackerels in different provincial dialects, having some mixture of gold or red in the plumage and white fluff, often extremely pretty. Legs white or yellow.

The brown-breasted yellow birchen
Male plumage: Breast reddish brown; hackles and saddle straw, striped birchen brown; shoulders old gold or birchen; wing bar and closed secondaries brown; tail brown or bronze-black.

Female plumage: Yellow-brown, with grey hackle and robin breast.

In both sexes: Fluff light grey. Beak, legs and eyes yellow.

The hennie
Male and female plumage: Hencocks should in their plumage resemble hens as closely as possible. They should have their hackle and saddle

Old English Game

feathers rounded and the tail coverts hen-like, and not have much sheen on their feathers. This breed often runs large and reachy, which is one of its characteristics. The two centre tail feathers should be straight.

The muff and tassel
Male and female plumage: Both muffs and tassels, or topins, are recognized by the Club, there being famous strains of both, though now scarce. Tassels vary from a few long feathers (or lark tops) behind the comb to a good-sized bunch. They also occur in some strains of hennies. Muffs of the old breed are stronger, heavier boned birds than the males bred today, and are rather loose in feather.

Notes:
(1) It is desirable that the toe-nails should match the legs and beak in colour in all Game Fowl, and that legs, eyes, beak and face match the male in all Game females.
(2) White or yellow legged birds may have white feathers in wings and tail.
(3) The fancier, when he speaks of a brown-red, means the streaky-breasted orange-red; and when talking of a black-red, intends one to infer a black-breasted light red; while black-breasted dark greys are erroneously called 'Birchens', although they have no birchen colour in them.

Weights

Male 2.50–2.55 kg (5½–5⅝ lb). It is not considered desirable to breed males over 2.70 kg (6 lb).
Female 1.80–2.25 kg (4–5 lb)

Scale of points

Body (including breast, back and belly)	20
Handling (symmetry, cleverness, hardness of flesh and feathers, condition and constitution)	15
Head (including beak and eyes)	10
Neck	6
Shanks, spurs and feet	10
Plumage and colour	9
Thighs	8
Wings	7
Tail	6
Carriage, action and activity	9
	100

Serious defects

Thin thighs or neck. Flat sided. Deep keel. Pointed, crooked or indented breast bone. Thick insteps or toes. Duck feet. Straight or stork legs. In-knees. Soft flesh. Broken, soft or rotten plumage. Bad carriage or action. Any indication of weakness of constitution.

Old English Game Bantams

This standard is compiled from that of the O.E.G. Bantam Club, and follows the Oxford ideal. Other standards exist, but essential differences are slight. Chief variations are in methods of interpretation. O.E.G. bantams are of comparatively recent creation. They were evolved largely from the common crossbred bantam of the countryside. Probably there is very little large breed blood in them.

In the large breed it is usually agreed that a good Game bird cannot be a bad colour. This remark does not apply to bantams, which are show birds only, colour playing a very important part. Nevertheless, the ideal is that the bantam should be a true miniature of the national fighting Game – though this is seldom the case.

General characteristics: male

Carriage: Bold, sprightly, defiant and proud, active on feet, agile, quick in movement, ready for any emergency.

Type: Body short. Back flat, broad at shoulders, tapering to tail. Breast broad, full and prominent, with strong pectoral muscles and breast bone not deep or pointed. Belly small and tight. Wings strong and powerful and carried well up. Tail large, carried upwards and spread, quills and main feathers large and strong.

Head: Small and tapering, skin of face and throat flexible and loose. Beak big and boxing (the upper mandible shutting tightly over the lower), crooked or hawk-like, pointed, strong at base. Eyes large, bold, fiery and alike in colour. Comb single, small, upright and thin. Ear-lobes and wattles fine, small and thin. (It is customary to dub males. Lobes and wattles are trimmed clean, but the comb is not dubbed so closely as in Modern Game.)

Neck: Large boned, round, strong and fairly long, neck hackle long and wiry and covering the shoulders.

Legs and feet: Legs strong, clean boned, sinewy, close scaled, not fat and gummy, not stiffly upright, not too wide apart, with a good angle at the hock. Thighs short, round and muscular, following line of body, slightly curved. Toes, four, long, straight and tapering, with long curved nails: hind toe of good length and strength, straight and firm on ground. Spurs hard and fine, set low.

Plumage: Hard, resilient, smooth and glossy, without much fluff.

Handling: Well balanced, mellow, firm and corky, hard yet light fleshed, with strong contraction of wings and thighs to body.

Female

The general characteristics are similar to those of the male, allowing for the natural sexual differences. The tail is inclined to fan-shape and carried well up.

Old English Game Bantams

Colour

Colours are very numerous, the most popular being spangles, black-reds (wheaten bred and partridge bred), duckwings, brown-reds, self blacks and blues, furnaces, creles and greys. Most of these main colours follow the usual colour-pattern applicable to the variety.

Furnaces are black with brassy hackles, wing shoulders and backs in males; and black with greyish brown streaked breasts and wings in females. Creles are barred varieties of other colours, distinct from cuckoos, which are plain barred grey. In black-reds, the wheaten bred is brighter in colour and more popular than partridge bred. Off colours are very numerous, but not so popular as in the large breed. Piles and whites are seldom seen. Hennies (hen-feathered males), muffs and tassels (beards and top-knots) are recognized sub-varieties in all colours.

Colour of face, beak, eyes, shanks and toe-nails varies with the particular colour-variety; from red face, red eyes, white legs, toe-nails and beak in spangles to deep crimson, purple or black faces, black eyes, black legs and feet in brown-reds and similar dark colours. Willow, olive and yellow legs are frequent and permissible in various sub-varieties.

Some of the more popular colour-varieties, with the names usually applied to them by showmen, are given below.

The black-red (partridge bred)
Male plumage: Cap dark red, neck hackle and saddle hackle dark rich red shading to deep orange. Back and saddle deep crimson. Breast, thighs, belly and tail black. Wing bars steel blue, secondaries showing bay when closed.
Female plumage: Hackle golden colour lightly striped with black. Breast salmon to robin red, shading off to ash grey on belly and thighs. Back, shoulders and wings even partridge brown stippled with fine darker markings. Primaries and tail dark partridge to black.

In both sexes: Eyes dark red, face blood red. Legs and beak willow, yellow, white or olive. Dark legged birds should have grey fluff, white or yellow legged birds should have light fluff.

The black-red (wheaten bred)
Male plumage: Cap bright red, neck hackle bright orange shading off to bright lemon. Saddle hackle similar. Back and saddle bright crimson. Breast, thighs, belly and tail black. Otherwise similar to partridge bred male but much brighter in colour.
Female plumage: Wheaten colour (a pale creamy tint resembling wheat) with clear red hackle. Tail and primaries nearly black, the breast and underparts a very delicate creamy self colour. Colour varies from very pale to a dark or red wheaten (the colour of red wheat) with dark red hackles and black tail feather.

In both sexes: Eyes fiery red, face bright red. Legs and beak white occasionally yellow. Fluff white.

The yellow (golden) duckwing
Male plumage: Face red. Hackle yellow, saddle straw colour, shoulders golden. Wing bars steel blue. Secondaries white when closed. Remaining plumage black.

Old English Game Bantams

Old English Game bantams
Black male
Wheaten female

Old English Game Bantams

Old English Game bantams
Duckwing male
Spangled female

Female plumage: Hackle white lightly striped with black. Body and wings deep silvery grey. Breast deep salmon shading off to ash grey on thighs. Primaries and tail nearly black.

In both sexes: Eyes pearl or red, beak and legs yellow, white or dark. Fluff light grey.

The silver duckwing
Male plumage: Breast, thighs, belly and tail black. Wing bars steel blue. Hackles, saddle, shoulders clear silvery white. Secondaries white when closed.
Female plumage: White hackle lightly striped with black. Body and wings even silvery grey. Breast pale salmon. Primaries and tail nearly black.

In both sexes: Eyes pearl or red. Beak and legs white or dark. Eye colour light or dark according to leg colour.

The brown-red
Male plumage: Hackles, back and wing bow orange to lemon. Remainder green-black, the breast feathers edged with orange as far down as top of thighs. Colour generally deeper than in Modern Game.
Female plumage: Cap orange, neck hackle light orange. Remainder green-black, breast feathers delicately laced with orange as in male. Back free from lacing and shoulders free from ticking.

In both sexes: Eyes black, beak and legs black, faces deep purple-black.

The spangled
Male plumage: Closely resembles the partridge bred black-red variety but with ends of feathers regularly tipped with small white spangles. The more regularly spangles are distributed the better. Neck and saddle hackle should be similarly tipped.
Female plumage: Closely resembling the partridge bred black-red female but each feather finished with a small white spangle tip – the more evenly distributed the better.

In both sexes: Eyes dark red, face blood red. Legs and beak usually white, occasionally yellow.
Note: In O.E.G. bantam spangles are always of partridge bred colouring. Spangles are not known in wheaten bred black-reds.

The self coloured: *Blacks* and *blues* are very popular. They should be free from white or coloured feathers, but may have red eyes, red faces and white legs, or black eyes, purple-black faces and black legs and beaks. *Whites* are seldom seen in bantams. They should be free from coloured feathers, with eyes red, faces red, legs and beaks white or yellow.

Although colour is less important in Old English Game than in Moderns, general colour descriptions in the main varieties are similar.

Weights (suggested)
Male   620–740 g (22–26 oz)
Female 510–620 g (18–22 oz)

Scale of points

Head, beak and eyes	10
Body, breast, back and belly	20
Tail	6
Shanks, spurs and feet	10
Handling, hardness, condition and constitution	15
Neck	6
Wings	7
Thighs	8
Plumage and colour	9
Carriage and action	9
	100

Serious defects

Thin thighs or neck. Deep keel. Pointed, crooked or indented breast bone. Flat sides. Stork legs. Duckfeet. In-knees. Soft flesh. Thick insteps or toes. Bad carriage. Bad action. Broken, soft or rotten plumage. Indications of weak constitution.

Old English Pheasant Fowl

Origin: Great Britain
Classification: Light
Egg colour: White

This breed was given its name of Old English Pheasant Fowl about 1914, previous to which it had been called the Yorkshire Pheasant, Golden Pheasant and also the Old-fashioned Pheasant. That it is a very old English breed is certain. Some Northern breeders retained their strains as Yorkshire Pheasant Fowls until the present tag of 'Old English' was brought officially into use. It has a meaty breast for a light breed, and has always been popular with farmers.

General characteristics: male

Carriage: Alert and active.

Type: Body rather long, deep and round with prominent shoulders. Tail flowing and set well back.

Head: Fine. Beak of medium size. Eyes bright and prominent. Comb rose, moderate in size, not impeding either sight or breathing, fine texture, evenly set, rather square front, the top flat and with plenty of work,

Old English Pheasant Fowl

Old English Pheasant Fowl large fowl Female

tapering to a single leader (or spike) at the back, which should gracefully curve downwards, following the neck line but quite free from it. Face and wattles smooth, free from coarseness or wrinkles. Ear-lobes medium size, oval or almond shape, smooth.

Neck: Graceful

Legs and feet: Legs of medium length, well apart, neither coarse nor too fine. Shanks free from feathers. Toes, four, well spread.

Female

With the exception of the tail (moderately whipped) the general characteristics are similar to those of the male, allowing for the natural sexual differences.

Colour

The gold
Male plumage: Ground colour bright rich bay; back rich mahogany red; lacing, bars, striping, tipping and tail beetle green-black. Hackles striped and slightly tipped; saddle a slightly deeper shade than neck. Breast laced. Wing bars (two) marked.
Female plumage: Ground colour bright rich bay; striping, tipping, spangling and tail beetle green-black. Neck with heavy stripe down centre of

Old English Pheasant Fowl

Old English Pheasant Fowl large fowl Male

each feather. Wing bars crescent spangling; even and well-marked bars a point of great beauty. Tail with slight edging of ground colour carried up from the base along the upper edge of the tail. Remainder, each feather tipped with a crescent-shaped spangle. Shafts of all feathers bay.

The silver
Male and female plumage: White with beetle green-black markings.

In both sexes and colours: Beak horn. Eyes fiery red. Comb bright rich red. Face and wattles red. Ear-lobes white. Legs and feet slate blue.

Weights

Cock 2.70–3.20 kg (6–7 lb); cockerel 2.50–2.70 kg (5½–6 lb)
Hen 2.25–2.70 kg (5–6 lb); pullet 2.00–2.25 kg (4½–5 lb)

Scale of points

Type (including legs)	20
Head (comb 15, lobes 5, other points 5)	25
Markings	20
Ground colour of body	15
Plumage and flow of feather	10
Size	5
Condition	5
	100

Serious defects

Comb single or over either side. Blushed lobes. Sooty hackles. Definitely black breast in male. Superfine bone. Lack of size. Squirrel tail. Any deformity. Any other defects which would affect health, hardiness, productivity or activity up to 20 points.

Orloff

Large fowl

Origin: Russia
Classification: Heavy
Egg colour: Tinted

The Orloff of Russian origin first appeared in this country around 1900. The Malay and some bearded Continental breeds are obviously in their make-up. Their gloomy vindictive expression and thick necks make them distinctive from any other breed. All four standard colours appear at most shows which cater for them.

General characteristics: male

Carriage: Upright with slightly sloping back.

Type: Body broad and fairly long; flat, slightly sloping back; breast rather full and prominent. Closely carried wings of moderate length; tail of medium size with fairly narrow sickles; carriage rather low but slightly above horizontal.

Head: Skull wide, or medium size. Beak short, stout and well hooked. Eyes full, and deeply set under well-projecting (beetle) eyebrows, giving a gloomy vindictive expression. Comb low and flat, shaped somewhat like a

Orloff

Orloff large fowl
White male

raspberry cut through its axis (lengthwise), covered with small protuberances mingled with small bristle-like feathers, which peculiarity is particularly noticeable in the female. Face muffled, beard and whiskers well developed. Ear-lobes very small, hidden under the muffles. Wattles small, and show only in the male.

Neck: Fairly long and erect, very heavily covered with hackle, the feathers very full at the top but so close at the base of the neck as to appear thin there, and forming a distinct angle with the back.

Legs and feet: Moderately long and stout. Thighs muscular and well apart. Shanks round and finely scaled. Toes, four, long and well spread.

Female

With the exception of the muffling (which is more developed) and the tail (comparatively long) the general characteristics are similar to those of the male, allowing for the natural sexual differences.

Colour

The black
Male and female plumage: Solid black to the skin from head to tail, with a beetle green sheen.

Orloff

The mahogany
Male plumage: Beard and whiskers a mixture of black, mahogany and grey, grey preponderating. Neck hackle rich dark orange, darkest at the crown and showing very slight black stripes at the base only. Saddle rich mahogany shading to deep orange. Wings rich deep mahogany with a strongly defined green-black sheen.
Female plumage: Mufflings as in the male. Hackle mahogany, the lower feathers showing black striping. Tail mainly black. Remainder rich dark mahogany uniformly peppered with black, the entire absence of black, or heavy and irregular black splashes undesirable.

The spangled
Male plumage: Hackles rich orange, with white tips to as many feathers as possible. Back rich mahogany. Wings rich mahogany with black bar showing green or purple sheen, and white flights. Breast solid black with white tips, blotchiness or washiness undesirable. Tail green-black.
Female plumage: Light mahogany with white tips, the spangling to be as uniform as possible.

The white
Male and female plumage: Lustrous white from head to tail.

In both sexes and all colours: Beak yellow, with a thin rose-tinted skin at base of beak and nostrils. Eyes red or amber. Comb, face, ear-lobes and wattles red. Legs rich yellow.

Weights

Cock 3.60 kg (8 lb); cockerel 3.20 kg (7 lb)
Hen 2.70 kg (6 lb); pullet 2.25 kg (5 lb)

Scale of points

Type and carriage	25
Colour	20
Comb and other head points Muffling	35
Legs	10
Condition	10
	100

Serious defects

Absence of beard and muffling, and puffed hackle. Legs other than yellow. Comb of any other form than as described. Weak, deformed or diseased specimens.

Disqualifications

In this breed the colour is of secondary importance and is a deciding point only in close competition. The main characteristics of the Orloff are its peculiarities of shape, comb, head and carriage and judges are earnestly requested to bear this in mind when awarding prizes. Slight feathering or down between the toes is not to constitute a disqualification.

Bantams

Orloff bantams follow the large fowl standard.

Orpington

Large fowl

Origin: Great Britain
Classification: Heavy
Egg colour: Brown

In the Orpington we have an English breed named after the village in Kent where the originator, William Cook, had his farm. He introduced the black variety in 1886, the white in 1889, and the buff in 1894. Within five years of the original black Orpington being introduced exhibition breeders were crossing Langshan and Cochin and exhibiting the offspring as black Orpingtons, the birds fetching high prices, and attracting many for their immense size. But this crossing at once turned a dual-purpose breed into one solely for show purposes, and it has remained so till today. A late introduction, the Jubilee Orpington, is now never seen.

General characteristics: male

Carriage: Bold, upright and graceful; that of an active fowl.

Type: Body deep, broad and cobby. Back nicely curved with a somewhat short, concave outline. Saddle wide and slightly rising, with full hackle. Breast broad, deep and well rounded, not flat. Wings small, nicely formed and carried closely to the body, the ends almost hidden by the saddle hackle. Tail rather short, compact, flowing and high, but by no means a squirrel tail.

Head: Small and neat, fairly full over the eyes. Beak strong and nicely curved. Eyes large and bold. Comb single, small, firmly set on head, evenly serrated and free from side sprigs. In the black variety, comb may be single or rose, the latter small, straight and firm, full of fine work or small spikes, level on top (not hollow in centre), narrowing behind to a distinct peak lying well down to the head (not sticking up). Face smooth. Wattles of medium length, rather oblong and nicely rounded at the bottom. Ear-lobes small and elongated.

Neck: Of medium length, curved, compact and with full hackle.

Orpington

**Orpington large fowl
Buff male
Buff female**

Orpington

Legs and feet: Legs short and strong, the thighs almost hidden by the body feathers, well set apart. Toes, four, straight and well spread.

Plumage: Fairly profuse but close, not soft, loose and fluffy as in the Cochin, or close and hard as in the Game Fowl.

Handling: Firm.

Female

The general characteristics are similar to those of the male. Her cushion should be wide but almost flat, and slightly rising to the tail, sufficient to give the back a graceful appearance with an outline approaching concave.

Colour

The blue
Male plumage: Hackles, saddle, wing bow, back and tail dark slate blue. Remainder medium slate blue, each feather to show lacing of darker shade as on back.
Female plumage: Medium slate blue, laced with darker shade all through, except head and neck, dark slate blue.

In both sexes: Beak black. Eyes black or very dark brown, black preferred. Comb, face, wattles and ear-lobes bright red. Legs and feet black or blue. Toe-nails white.

The black
Male and female plumage: Black with a green sheen.

In both sexes: Beak, etc. as in the blue. Soles of feet white.

The buff
Male and female plumage: Clear, even buff throughout to the skin.

In both sexes: Beak white or horn. Eyes red or orange colour. Comb, face, ear-lobes and wattles bright red. Legs, feet and toe-nails white. Skin white.

The white
Male and female plumage: Pure snow white.

In both sexes: Beak, legs, feet and skin white. Eyes, face, ear-lobes and wattles red.

Weights

The blue
Male   4.55–6.35 kg (10–14 lb) when fully matured.
Female 3.40–4.80 kg (7½–10½ lb)

The black
Male   4.55 kg (10 lb)
Female 3.60 kg (8 lb)

The buff and white
Matured cockerel 3.60–4.55 kg (8–10 lb)
Female   2.70–3.60 kg (6–8 lb)
  Old birds sometimes a little heavier.

Orpington

**Orpington large fowl
Black male
White female**

Orpington

Scale of points

The blue

Type	25
Size (with utility qualities)	20
Head	10
Legs and feet	10
Colour and plumage	25
Condition	10
	100

The black

Type (shape) (body 15, breast 10, saddle 5)	30
Size	10
Carriage	10
Head (skull 5, comb 7, face 5, eyes 5, beak 3)	25
Skin	5
Legs and feet	5
Plumage and condition	10
Tail	5
	100

The buff and white

Type	30
Size	10
Head	15
Legs and feet	10
Colour	20
Condition	15
	100

Serious defects

Side spikes on comb. White in ear-lobes. Feather or fluff on shanks or feet. Long legs. Any deformity. Yellow skin or yellow on the shanks or feet of any variety. Any yellow or sappiness in the white. Coarseness in head, legs or feathers of the buff.

Disqualifications

Trimming or faking.

Orpington

Orpington bantam Black male

Bantams

Orpington bantams are miniatures of their large fowl counterparts and the standard for those should be followed.

Weights

Male   910–1020 g (32–36 oz)
Female 790–910 g (28–32 oz)

Scale of points

	Black	Blue	Buff and white
Type, carriage and feather	35	30	30
Colour and undercolour	15	30	20
Head and eyes	20	15	15
Legs, feet and skin	15	10	10
Size and condition	15	15	25
	100	100	100

Pekin Bantams

Origin: Asia
Classification: True bantam
Egg colour: White or cream

This is a genuine bantam breed, very old and having no real relationship to the large breed of Cochins. It was imported from Pekin in the middle of the 19th century, hence its name. In recent years new colours have been added to the standard.

General characteristics: male

Carriage: Bold, rather forward and low, the head very little higher than the tail.

Type: Body short and broad. Back short, increasing in breadth to the saddle, which should be very full, rising well from between the shoulders and furnished with long soft feathers. Breast deep and full. Wings short, tightly tucked up, the ends hidden by saddle hackle. Tail very short and full, soft and without hard quill feathers, with abundant coverts almost hiding main tail feathers, the whole forming one unbroken duplex curve with the back and saddle. General type as much like a ball as possible.

Head: Skull small and fine. Beak rather short, stout, slightly curved. Eyes large and bright. Comb single, small, firm, perfectly straight and erect, well serrated, curved from front to back. Face smooth and fine, ear-lobes smooth and fine, preferably nearly as long as the wattles, which are long, ample, smooth and rounded.

Neck: Short, carried forward, with abundant long hackle reaching well down the back.

Legs and feet: Legs short and well apart. Stout thighs hidden by plentiful fluff. Hocks completely covered with soft feathers curling round the joints (stiff feathers forming 'vulture hocks' are objectionable but not a disqualification). Shanks short and thick, abundantly covered with soft outstanding feathers. Toes, four, strong and straight, the middle and outer toes plentifully covered with soft feathers to their tips.

Plumage: Very abundant, long and wide, quite soft with very full fluff.

Female

With the exception of the back (rising into a very full and round cushion) the general characteristics are similar to those of the male, allowing for the natural sexual differences.

Colour

The black
Male and female plumage: Rich sound black with lustrous beetle green sheen throughout, free of white or coloured feathers. (*Note:* Some light

Pekin Bantams

Pekin bantams
White male
Mottled female

undercolour in adult males is permissible as long as it does not show through.)

The blue
Male and female plumage: A rich pale blue (pigeon blue preferred) free from lacing, but with rich dark blue hackles, back and tail in the male.

The buff
Male and female plumage: Sound buff, of a perfectly even shade throughout, quite sound to roots of feathers, and free from black, white or bronze feathers. The exact shade of buff is not material so long as it is level throughout and free from shaftiness, mealiness or lacing. (*Note:* A pale 'lemon buff' is usually preferred in the show-pen.)

The cuckoo
Male and female plumage: Evenly barred with dark slate on light French grey ground colour.

The mottled
Male and female plumage: Evenly mottled with white at the tip of each feather on a rich black with beetle green sheen.

The barred
Male and female plumage: Each feather barred across with black bands, having a beetle green sheen on a white background. The bands or barring to be equal proportions of black and white. The colours to be sharply defined and not blurred or shaded off. Barring should continue through the shaft and into the underfluff, and each feather must finish with a black tip. Plumage should present a bluish steely appearance free from brassiness and of a uniform shade throughout.

In both sexes: Eyes red orange. Legs and feet yellow.

The columbian
Male and female plumage: Pearl white with black markings. Head and neck hackle white with dense black stripe down middle of each feather, free from black edgings or black tips. Saddle pearl white. Tail feathers and tail coverts glossy green-black, the coverts laced or not with white. Primaries black, or black edged with white. Secondaries black on inner edge, white outer. Remainder of plumage entirely white, of pearl grey shade, free from ticking. Undercolour either slate, blue-white or white.

The lavender
Male and female plumage: The lavender is not a lighter shade of the blue Pekin. It is different genetically and is of a lighter more silver tint without the darker shade associated with the normal blue. The silver tint is most obvious in the neck and saddle hackle feathers of the male.

In both sexes: Beak yellow or horn. Eyes orange red-brown. Legs and toes deep yellow.

The partridge
Male plumage: Head dark orange red, neck hackle bright orange or golden red, becoming lighter towards the shoulders and preferably shading off as near lemon colour as possible, each feather distinctly striped down the middle with black, and free from shaftiness, black tipping or black

fringe. Saddle hackle to resemble neck hackle as nearly as possible. Breast, thighs, underparts, tail, coverts, wing butts and foot feather, hock feather and fluff lustrous green-black, free from grey, rust or white. Back, shoulder coverts and wing bow rich crimson. Primaries black, free from white or grizzle. Secondaries black inner web, bay outer, showing a distinct wing bay when closed.

Female plumage: Head and neck hackle light gold or straw, each feather distinctly striped down middle with black. Remainder clear light partridge brown, finely and evenly pencilled all over with concentric rings of dark shade (preferably glossy green-black). The whole of uniform shade and marking, and the ground colour of the soft brown shade frequently described as the colour of a dead oak leaf, with three concentric rings of pencilling or more over as much of the plumage as possible.

The white

Male and female plumage: Pure snow white, free from cream or yellow tinge, or black splashes or peppering.

In both sexes and all colours: Beak yellow, but in dark colours may be shaded with black or horn. Eyes red, orange or yellow – red preferred. Comb, face, wattles and ear-lobes bright red. Legs and feet yellow. (Dark legs are permissible in blacks if the soles of the feet and back of shanks are yellow.)

Weights

British Bantam Association
Male   680–790 g (24–28 oz)
Female 570–680 g (20–24 oz)

Club standard
Male   680 g (24 oz) max.
Female 570 g (20 oz) max.

Scale of points

Colour and markings	20
Fluff and cushion	15
Leg and foot feather	10
Size and weight	10
Type and carriage	15
Head	10
Length of shank	10
Condition	10
	100

Serious defects

Twisted or drooping comb. Slipped wings. Legs other than yellow (except for blacks). Eyes other than red, orange or yellow. Any deformity.

Plymouth Rock

Large fowl

Origin: America
Classification: Heavy
Egg colour: Tinted

Specimens of the barred Plymouth Rock were first exhibited in America in 1869, and stock reached here in 1871. The white and black varieties came as sports. About 1890 the buff was exhibited in America and in England. The barred Rock came to us as a dual-purpose breed, but was developed to an exhibition ideal in which body size and frontal development were neglected in order to secure long narrow finely-barred feathers. With the introduction of sex-linkage between the black Leghorn and barred Rock for commercial purposes, utility breeders made use of the Canadian barred Rock, a bird with roomy body, full breast, lower on the leg but coarser in barring.

General characteristics: male

Carriage: Alert, upright with bold appearance, well balanced and free from stiltiness.

Type: Body large, deep and compact, evenly balanced and symmetrical, broad, the keel bone long and straight. Back broad and of medium length, saddle hackle of good length and abundant. Breast broad and well rounded. Wings of medium size, carried well up, bow and tip covered by breast and saddle feathers respectively; flights carried horizontally. Tail medium size, rising slightly from the saddle to be carried neatly and not to be fan, squirrel, or wry tail, sickles medium length and nicely curved, coverts sufficiently abundant to cover the stiff feathers.

Head: Of medium size, strong and carried well up. Beak short, stout and slightly curved. Eyes large, bright and prominent. Comb single, medium in size, straight and erect with well-defined serrations, smooth and of fine texture, free from side sprigs and thumb marks. Face smooth. Ear-lobes well developed, pendent, and of fine texture. Wattles moderately rounded and of equal length, to correspond with size of comb, smooth and of fine texture.

Neck: Of medium length, slightly curved, a full hackle flowing over the shoulders.

Legs and feet: Legs wide apart. Thighs large and of medium length. Shanks medium length, stout, well rounded, smooth and free from feathers. Toes, four, strong and perfectly straight, well spread and of medium length.

Fluff: Moderately full, carried closely to the body and of good texture.

Skin: Silky and fine in texture.

Plymouth Rock

Plymouth Rock large fowl
Barred male
Barred female

Plymouth Rock

Female

The general characteristics are similar to those of the male, allowing for the natural sexual differences, except that comb, ear-lobes and wattles are smaller, the neck is of medium length, carried slightly forward, and the tail is small and compact, carried well back.

Colour

The barred
Male and female plumage: Ground colour, white with bluish tinge, barred with black of a beetle green sheen, the bars to be straight, moderately narrow, of equal breadth and sharply defined, to continue through the shafts of the feathers. Every feather to finish with a black tip. The fluff, or undercolour, to be also barred. The neck and saddle hackles, wing bow and tail to correspond with the rest of the body, presenting a uniformity of colour throughout.

The black
Male and female plumage: Black with a bettle green sheen

The buff
Male and female plumage: Clear, sound, even golden buff throughout to the skin. The tail clear buff to harmonize with the body colour, and the undercolour (or fluff) and the quill of the feather also to harmonize with the surface colour. The male of more brilliant lustre than the female.

The columbian
Male and female plumage: Colour and markings as in light Sussex.

The white
Male and female plumage: Pure snow white, any straw tinge to be avoided.

In both sexes: Beak yellow. Eyes rich bay. Face, comb, ear-lobes and wattles red. Legs and feet yellow.

Weights:

The buff
Cock 3.85–4.80 kg (8½–10½ lb); cockerel 3.20–3.85 kg (7–8½ lb)
Hen 2.70–3.40 kg (6–7½ lb); pullet 2.25–2.95 kg (5–6½ lb)

Other varieties
Cock 4.10–4.55 kg (9–10 lb); cockerel 3.40–4.30 kg (7½–9½ lb)
Hen 3.20–3.60 kg (7–8 lb); pullet 2.70–3.20 kg (6–7 lb)

Scale of points

The barred
Type	30
Barring	20
Colour	15
Size	10

(*continued*)

Plymouth Rock

Plymouth Rock large fowl
Buff male
Buff female

Plymouth Rock

(Scale of points – *continued*)

Legs and feet	10
Condition	10
Head	5
	100

The buff

Type (symmetry), shape, size and carriage	30
Colour (general)	20
Quality and texture (general)	15
Condition and fitness	15
Head and comb	10
Eye colour	5
Legs and feet	5
	100

Other varieties

Type	30
Colour	30
Head and eyes	10
Legs and feet	10
Condition	10
Quality and texture	10
	100

Serious defects

The slightest fluff or feather on shanks or feet, or unmistakable signs of feathers having been plucked from them. Legs other than yellow. White in ear-lobes. In the barred, any feathers of any colour foreign to the variety, black feathers excepted; also lopped or rose comb, decidedly wry tail; crooked back, more than four toes, and entire absence of main tail feathers. Other than black feathers in the black. Mealiness, or any black or white in wing or white in tail, spotted hackle, and in the male a spotted saddle and in the female a spotted cushion in the buff. Any coloured feathers in the white.

Disqualifications

Trimming, faking and any bodily deformity. Split wing, slipped wing and non-growth of secondaries.

Bantams

Plymouth Rock bantams are miniatures of their large fowl counterparts and the standards for large fowl should be used. Buff, barred and partridge are the three colours standardized for bantams although blacks and whites have been seen.

Plymouth Rock

Plymouth Rock bantam Buff male

Colour

The partridge
Male plumage: Head bright red. Neck hackles: web of feather, solid, lustrous, greenish black, of moderate width, with a narrow edging of a medium shade of rich brilliant red, uniform in width, extending around point of feather; shaft black; plumage on front of neck black, as clear as possible of red. Wing fronts black; bows a medium shade of rich brilliant red; coverts lustrous, greenish black, forming a well-defined bar of this colour across wing when folded; primaries black, lower edges reddish bay; secondaries black, outside webs reddish bay, terminating with greenish black at the end of each feather, the secondaries when folded forming a reddish bay wing bay between the wing bar and tips of secondary feathers. Back and saddle a medium shade of rich, brilliant red, with lustrous, greenish black stripe down the middle of each feather, same as in hackle. A slight shafting of rich red is permissible. Tail black; sickles and smaller sickles lustrous, greenish black; coverts lustrous, greenish black, edged with a medium shade of rich, brilliant red. Body black; lower feathers slightly tinged with red; fluff black, slightly tinged with red. Breast lustrous black (slight tinge of red allowed). Lower thighs black. Undercolour of all sections slate.

Plymouth Rock

Plymouth Rock bantam Barred female

Female plumage: Head deep reddish bay. Neck reddish bay, centre portion of feathers black, slightly pencilled, with deep reddish bay; feathers on front of neck same as breast. Wing shoulders, bows and coverts deep reddish bay with distinct pencillings of black, outlines of which conform to shape of feathers; primaries black with edging of deep reddish bay on outer webs; secondaries, inner web black, outer web deep reddish bay with distinct pencillings of black extending around outer edge of feathers. Back deep reddish bay with distinct pencillings of black, the outlines of which conform to shape of feathers. Tail black, the two top feathers pencilled with deep reddish bay on upper edge; coverts deep reddish bay pencilled with black. Body deep reddish bay pencilled with black; fluff deep reddish bay. Breast deep reddish bay with distinct pencillings of black, the outlines of which conform to shape of feathers. Lower thighs deep reddish bay pencilled with black. Undercolour slate.
Note: Each feather in back, breast, body, wing bows and thighs to have three or more distinct pencillings.

In both sexes: Beak yellow. Eyes reddish bay. Legs and feet yellow. Comb, face, wattles and ear-lobes bright red.

Weights

British Bantam Association
Male 740–850 g (26–30 oz)
Female 620–740 g (22–26 oz)

The standards of the Buff Rock Blub and the Barred Rock Bantam Club were respectively slightly below and above these figures.

Scale of points

The barred
Type	20
Colour	20
Barring	20
Legs and feet	10
Head	5
Tail	5
Size	10
Condition	10
	100

The buff
Type	20
Carriage and size	10
Colour	20
Quality and texture	15
Head	10
Eye colour	5
Legs and feet	5
Condition	15
	100

The partridge
Type	30
Colour and pencilling	30
Head and eyes	10
Legs and feet	10
Condition	10
Quality and texture	10
	100

Poland

Large fowl

Origin: Poland
Classification: Light
Egg colour: White

That the Poland is a very old breed goes without saying, although its ancestry is none too clear. Many connect it with the breed named the Paduan or Patavinian fowl, although this original example is illustrated without muff or beard. Polish (gold or silver spangled, black or white) had

Poland large fowl
White-crested
black female

a classification at the first poultry show in London in 1845, and was standardized in the first Book of Standards in 1865, with white-crested black, golden and silver varieties included. The white-crested (black or blue) varieties are without muffling, while the others have muffs.

General characteristics: male

Carriage: Sprightly and erect.

Type: Back fairly long, flat and tapering to the tail. Breast full and round. Flanks deep. Shoulders wide. Wings large and closely carried. Tail full, neatly spread, and carried somewhat low, not perpendicularly, the sickles and coverts abundant and well curved.

Head: Large, with a decidedly pronounced protuberance on top, and crested. Crest large, full, circular on top and free of any split or parting, high and smooth in front and compact in the centre, falling evenly with long untwisted or reverse-faced feathers far down the nape of the neck, and composed of feathers similar to those of the hackles. Beak of medium length, and having large nostrils rising above the curved line of the beak. Eyes large and full. Comb of horn type and very small if any (preference should be given to birds without a comb). Face smooth, without muffling in the white-crested black or blue varieties, and completely covered by muffling in the others. Muffling large, full and compact, fitting around to the back of the eyes and almost hiding the face. Ear-lobes very small and

Poland

round, quite invisible in the muffled varieties. Wattles rather large and long in the white-crested varieties; the others are without wattles.

Neck: Long, with abundant hackle coming well over the shoulders.

Legs and feet: Legs slender and fairly long, the shanks free of feathers. Toes, four, well spread.

Female

With the exception of the crest, which is of globular shape, the general characteristics are similar to those of the male, allowing for the natural sexual differences.

Colour

The chamois or white laced buff
Male plumage: Buff ground with white lacing. Crest white at the roots and tips, and as free as possible from whole white feathers. Muffling mottled or laced, not solid buff. Hackle tipped. Wing bar and secondaries laced, and primaries tipped. Tail, sickles and coverts laced.
Female plumage: Except that the wing primaries are tipped, the colours and markings, including the crest, are buff ground and white lacing.
   In both sexes: Beak dark blue or horn. Eyes, comb and face red. Ear-lobes blue-white. Legs and feet dark blue.

The gold
Male plumage: Golden bay ground with black markings. Crest black at roots and tip, and as free as possible of whole white feathers. Muffling mottled or laced, not solid black. Hackle tipped. Back and saddle distinctly laced or spangled at the tips. Breast, thighs, shoulders and wings laced except primaries (of wings) which are tipped. Tail laced, the ends of the sickles well splashed.
Female plumage: Golden bay ground with black lacing, each feather distinctly marked and as free as possible from splashes.
   In both sexes: Beak, etc. as in the chamois.

The silver
Male and female plumage: As in the gold, substituting silver as the ground colour.
   In both sexes: Beak, etc. as in the chamois.

The white
Male and female plumage: Pure white.
   In both sexes: Beak dark blue. Ear-lobes white. Other points as in the chamois.

The white-crested black
Male and female plumage: Rich metallic black, except the crest which is snow white, with a black band at base of crest in front.
   In both sexes: Beak, etc. as in the white.

Poland

The white-crested blue
Male and female plumage: Solid dark blue (self coloured). Crest snow white with a blue band at base of crest in front.

In both sexes: Beak, etc. as in the white.

The white-crested cuckoo
Male and female plumage: Even barring on all feathers from either a dark slate colour to a pale grey, but one shade only all over, except the crest, which is snow white with a cuckoo band at the base, and in front of the face.

In both sexes: Beak and legs white or showing a touch of horn or slate. If blue legs and beak could ever be perfected, this would be desirable. Eyes, wattles and face red. Skin colour white or showing blue tinge on those birds with darker legs and beak.

Serious defects: As stated for all other Polands, with exception to legs and beak colour. Self black or white feathers. Some white in sickles of second year males permissible.

Weights
Male   2.95 kg (6½ lb)
Female 2.25 kg (5 lb)

Scale of points
The white-crested
Type	5
Head (crest 30, comb and wattles 15)	45
Colour	30
Condition	15
Size	5
	100

Other varieties
Type	10
Head (crest 30, muffling 10)	40
Colour and markings	30
Condition	10
Size	10
	100

Serious defects

Split or twisted crest. Comb, if any, other than horn type. Absence of muffling in black, chamois, gold, silver, and white varieties. Legs other than blue or slate. Other than four toes on each foot. Any deformity. Absence of black or blue in the front of the crest in white-crested blacks and blues.

Poland

Poland bantams
White male
White-crested black female

Redcap

Bantams

Poland bantams follow the standard for their large fowl counterpart.

Weights

Male   680–790 g (24–28 oz)
Female 510–680 g (18–24 oz)
   The weights of some colours are often considerably higher.

Scale of points

The white-crested coloured
Crest	30
Head	15
Colour	25
Type	5
Size	15
Condition	10
	100

Other colours
Crest	30
Head and muffling	15
Colour and markings	25
Type	5
Size	15
Condition	10
	100

Redcap

Origin: Great Britain
Classification: Light
Egg colour: White

The Redcap has always been closely associated with Derbyshire and is known as the Derbyshire Redcap. It was a sturdy breed carrying excellent breast meat, and a good egg producer. Farmers used the males freely for crossing to produce layers. As with so many promising utility breeds, the Redcap was bred and exhibited as if its immense comb were all that mattered, head points claiming 45 of the 100 judging points.

General characteristics: male

Carriage: Graceful action and well balanced.

Redcap large fowl
Male

Type: Back broad, moderate in length falling slightly to tail and flat. Breast broad, full, and rounded. Wings moderate in length, neat and fitting closely to the body. Tail full and carried at an angle of about 60°; broad, long and well-arched sickles.

Head: Of medium length and broad. Beak medium in length. Eyes full and prominent. Comb rose with straight leader, full of fine work or spikes, free from hollow in centre, set straight on the head and carried well off the eyes and beak; size about 8.25 × 7.0 cm (3¼ × 2¾ in). Face smooth and of fine texture. Ear-lobes of medium size. Wattles of medium length, well rounded and fine in texture.

Neck: Of moderate length, nicely arched, and with full hackle.

Legs and feet: Legs straight and wide apart. Thighs short and well fleshed, shanks moderately long. Toes, four, well spread.

Female

The general characteristics are similar to those of the male, allowing for the natural sexual differences, with the exception of the comb which is about half the size of the male's. The tail is large and full, and, like that of the male, carried at an angle of about 60°.

Colour

Male plumage: Neck, hackle and saddle to harmonize. Each feather to have a red quill with beetle green webbing, very finely fringed and tipped with black (the feathers appear to be fringed with red, but if placed on paper they are seen to have a black fringe and tip; the fringe is almost as fine as a hair). Back rich red, tipped with black. Wing bows rich red. Coverts rich red, each feather ending with a black spangle, forming a black bar across the wing. Primaries and secondaries black on one side red on the other side, heavily tipped with black. Breast and underparts black. Tail and hangers black.

Female plumage: Hackle as in the male, but nut brown quill. Back and breast ground colour deep rich nut brown, free from smuttiness, each feather ending with a half-moon black spangle. The markings on breast, back and wings to be as uniform as possible. Wing primaries and secondaries, as in the male, wing coverts evenly spangled. Tail black.

In both sexes: Beak horn. Eyes red. Comb, face, ear-lobes and wattles bright red. Legs and feet lead colour.

Weights

Cock 2.70–2.95 kg (6–6½ lb); cockerel 2.50–2.70 kg (5½–6 lb)
Hen 2.25–2.50 kg (5–5½ lb); pullet 2.00–2.25 kg (4½–5 lb)

Scale of points

Size (style and shape)	10
Tail	5
Head points (comb 25, eyes 5, ear-lobes and wattles 15)	45
Legs and feet	5
Colour	25
Condition	10
	100

Serious defects (for which birds should be passed)

Comb over. White ear-lobes. Round back. Squirrel or wry tail. Feathers on legs. Legs other colour than lead. Other than four toes. Crooked breast. Coarseness. Excessive fat.

Rhode Island Red

Origin: America
Classification: Heavy
Egg colour: Light brown to brown

No breed made such a world progress in so short a time as this American breed. It was developed from Asiatic black-red fowls of the Shanghai, Malay, and Java types, bred on the farms of Rhode Island Province. Red

Javas were known there in 1860, and the original Rhode Island Red had a rose comb, although birds with single, probably from brown Leghorn crossings, and pea combs were also bred. The rose combs were first called American Reds and so standardized in 1905, a year after the single comb was accepted as the Rhode Island Red. The formation of the British Rhode Island Red Club took place in 1909, and the breed has been one of the most popular in this country for all purposes. Being a gold, males of the breed are utilized extensively in gold-silver sex-linked matings.

General characteristics: male

Carriage: Alert, active, and well balanced.

Type: Body deep, broad and long; the keel bone long, straight, and extending well forward and back, giving the body an oblong look. Back broad, long and in the main nearly horizontal, this being modified by slightly rising curves at hackle and lesser tail coverts. Saddle feathers of medium length and abundant. Breast broad, deep and carried in a line nearly perpendicular with the base of the beak; at least it should not be carried further back. Fluff moderately full, but with the feathers carried fairly close to the body; not a Cochin fluff. Wings of good size, well folded, and the flights carried horizontally. Tail of medium length, quite well spread, carried fairly well back, increasing the apparent length of the bird. Sickles of medium length, passing a little beyond the main tail feathers. Lesser sickles and tail coverts of medium length and fairly abundant.

Head: Of medium size, carried horizontally and slightly forward. Beak medium in length and slightly curved. Eyes full bright and prominent. Comb single or rose. The single of medium size, fine texture, set firmly in the head, perfectly straight and upright, with five even and well-defined serrations, those in front and rear smaller than the centre ones, of considerable breadth where it is fixed to the head. The rose of medium size, low, set firmly on the head, the top oval in shape, and the surface covered with small points, terminating in a small spike at the rear. The comb to conform to the general curve of the head. Face smooth and of fine texture. Ear-lobes fairly well developed. Wattles medium and equal in length, moderately rounded and of fine texture.

Neck: Of medium length, carried slightly forward, and covered with abundant hackle, flowing over the shoulders but not too loosely feathered.

Legs and feet: Legs well apart. Thighs large, of medium length and well covered with feathers. Shanks of medium length, well rounded and smooth. Toes, four, of medium length, straight, strong and well spread.

Female

The general characteristics are similar to those of the male, allowing for the natural sexual differences. The tail, however, should not form an apparent angle with the back, nor must it be met by a high rising cushion. It should be a little shorter than medium and quite well spread. Neck hackle should be sufficient, but not too coarse in feather. In the mature female the back

Rhode Island Red

Rhode Island Red
large fowl
Male
Female

would be described as broad, while in the pullet it would look somewhat narrower in proportion to the length of her body. The curve from the horizontal back to the hackle or tail should be moderate and gradual.

Colour

Male plumage: The neck red, harmonizing with back and breast. Wing primaries, the lower web black and the upper red; secondaries, the lower web red and the upper black; flight coverts black; wing bows and coverts red. Tail, main feathers, including the sickles, black or greenish black; coverts mainly black, but they may become russet or red as they approach the saddle.

The general surface of the plumage should be a rich brilliant red, except where black is specified. It should be free from shafting, mealy appearance or brassy effect. Absolute evenness of colour is desired.

The bird should be so brilliant in lustre as to have a glossed appearance. The undercolour and quill of the feather should be red or salmon. With the saddle parted, showing the undercolour at the base of the tail, the appearance should be red or salmon, not whitish or smoky. Black or white in the undercolour of any section is undesirable.

Female plumage: Neck hackle red, the tips of the lower feathers having black ticking, but not heavy lacing. The tail should be black or greenish black. In all sections of the wing the undercolour and quills of the feathers are as in the male. With the remainder of the plumage the surface should be a rich dark, even and lustrous red, but not as brilliant a lustre as in the male. It should be free from shafting or mealy appearance.

In both sexes, other thing being equal, the specimen having the richest undercolour shall receive the award.

In both sexes: Beak red-horn or yellow. Eyes red. Face, comb, wattles and ear-lobes bright red. Legs and feet yellow or red-horn.

Weights

Cock 3.85 kg (8½ lb); cockerel 3.60 kg (8 lb)
Hen 2.95 kg (6½ lb); pullet 2.50 kg (5½ lb)

Scale of points

Shape, size, carriage and symmetry	30
Colour (general)	20
Quality and texture (general)	15
Head and comb	10
Eye colour	10
Condition	10
Legs	5
	100

Serious defects

Feather or down on shanks or feet, or unmistakable indications of a feather having been plucked from them. Badly lopped comb, side sprig or sprigs on the single comb. Other than four toes. Entire absence of main tail feathers. Two absolutely white (so-called wall or fish) eyes. Squirrel or wry tail. A feather entirely white that shows in the outer plumage. An ear-lobe showing more than one-half of the surface permanently white. (This does not mean the pale ear-lobe, but the enamelled white.) Diseased specimens, crooked backs, deformed beaks, shanks and feet other than yellow or red-horn colour. A pendulous crop shall be cut hard. Coarseness. Toes not straight and well spread. Super-fineness. Under all disqualifying clauses, the specimen shall have the benefit of the doubt.

Robustness is of vital importance.

Rhode Island Red Bantams

There remain points of difference between the Rhode Island Red Club and the Rhode Island Red Bantam Club. The standard which follows is a compromise between the two.

General characteristics: male

Carriage: Active, alert and well balanced.

Type: Body broad and long, fair depth, distinctly oblong rather than square, with long straight keel bone extending well forward and back. Back broad, long and horizontal. Breast broad and full, carried nearly perpendicularly in line with base of beak. Wings large and well folded, with horizontal flights. Tail of moderate size (this is important), carried well back, the sickles a little longer than the main tail feathers, well spread and carried at a low angle (but not drooping) thus increasing the apparent length of the body. Feathering close, fluff moderately full.

Head: Medium size, carried slightly forward. Beak curved, moderate length. Eyes large, prominent and bright. Comb single, medium size, fine texture, erect, straight and firmly set, with five even serrations. (A rose-combed variety is standardized in the large breed, but seldom seen in bantams.) Face smooth. Ear-lobes fine and well developed. Wattles medium size and moderately rounded, of fine texture.

Neck: Medium length, profusely hackled but not loose feathered; carried slightly forward.

Legs and feet: Legs of medium length. Thighs large and well feathered; shanks well rounded and free of feathers. Toes, four, strong, straight and well spread.

Female

The general characteristics are similar to those of the male, allowing for the natural sexual differences.

Colour

Male plumage: Hackle red, without black markings, matching body colour. Wing primaries, lower web black, upper red; secondaries, lower web red, upper black; flight coverts black; wing bows and coverts red. Main tail and sickles black or green-black. Coverts mainly black, but red approaching the saddle. Remainder rich brilliant red, free from shaftiness, mealiness, peppering or 'ginger'. The bird should be of very brilliant lustre and having a glossed appearance. Undercolour and quill colour red or salmon, without smut or white. Black or white in undercolour is most undesirable; and (other things being equal) the richest red undercolour shall carry the award. (*Note:* The 'red' colour nowadays favoured is an extremely deep chocolate red, and though some breeders disagree with this description, few birds of lighter colour receive prizes.)

R.I.R. Club judges are now instructed to pass 'over-prepared' birds without comment, and to adhere strictly to the colour-standard, which calls for rich brilliant red.

Female plumage: Hackle red, tips of lower feathers having black ticking but not heavy markings. Tail black or green-black. Remainder generally as described for the male, except that the female will not be so lustrous.

In both sexes: Beak, legs and feet red-horn or yellow. Eyes red; comb, face, ear-lobes and wattles brilliant red.

Weights

Male   790–910 g (28–32 oz)
Female 680–790 g (24–28 oz)
   The Bantam Club favours lower weights.

Scale of points

Colour of plumage	25
Size	25
Type	15
Condition	10
Eye colour	9
Legs	6
Head and comb	10
	100

(*Note:* This scale of points is the one adopted by the Bantam Club. The scale followed by the Rhode Island Red Club approximates closely to that for larger Rhodes.)

Serious defects

Feather or down on shanks, or indications of plucking same. Lopped combs or side sprigs. Wall eyes. White showing in outer plumage. Lobes more than half white. Shanks and feet other than yellow or red-horn. Any deformity.

(*Note:* To these serious defects should be added Leghorn-type tails and the frizzled, Silkie type or otherwise defective feather often developed through concentration on lustrous dark plumage.)

Rosecomb Bantams

Origin: Great Britain
Classification: True bantam
Egg colour: White or cream

The Rosecomb bantam is a gem of show birds. In former days it achieved probably the highest pitch of artificial perfection ever achieved in exhibition birds.

General characteristics: male

Carriage: Cobby but not dumpy. The back should show one sweeping curve from neck to sickles.

Type: Body short and broad. Back short, shoulders broad and flat. Breast carried well up and forward, with a bold curve from wing bow to wing bow. Wings carried rather low, showing only front half of thighs. Wide flight feathers round ended and broad to ends. Stern flat, broad and thick (not running off to nothing at setting-on of tail), with abundant feather; the saddle hackle long and plentiful and extending from tail to middle of back. Tail carried well back, main feathers broad and overlapping neatly; the sickles being long, circled with a bold sweep, broad from base to rounded ends, main tail feathers not projecting beyond the sickles. Furnishing feathers plentiful, broad from base to end, round ended and uniformly curved with the sickles but hanging somewhat shorter; side hangers broad and long and with the hackles filling the space between stern and wing ends. All feather broad to ends.

Head: Short and broad. Beak stout and short. Comb rose, neat and long, with square well-filled front, set firmly, tapering to the setting-on of the spike or leader; top perfectly level and crowded with small round spikes. The leader stout at base, firm, long and perfectly straight, tapering to a fine point. Comb and leader rise slightly from front to rear in one line. Face of fine texture. Ear-lobes absolutely round, with rounded edges, of uniform thickness all over, not hollow or dished, firmly set on the face and kid-like in texture; not smaller than 1.88 cm (¾ in) or larger than 2.19 cm (⅞ in). Wattles round, neat and fine.

Neck: Rather short, well curved, with wide feathers, the hackle falling gracefully and plentifully over shoulders and wing bows and almost reaching the tail.

Legs and feet: Legs short. Thighs set well apart, stout at top and tapering to hocks. Shanks rather short, round, fine and free of feathers. Toes, four, straight and well spread.

Female

With the exception of the ear-lobes, which should not be larger than the now defunct silver threepenny piece (approx. 1.56 cm (⅝ in)), and the

Rosecomb Bantams

Rosecomb bantams
Black male
Black female

wings, which are not carried so low but are tucked up, the general characteristics are similar to those of the male, allowing for the natural sexual differences.

(*Note:* Standard sizes of ear-lobes are usually considerably exceeded in show specimens.)

Colour

The black
Male and female plumage: Black with brilliant green sheen from head to end of tail, the wing bar with extra bright green sheen. Tail feathers and sickles to be rich in green sheen.

In both sexes: Beak black, eyes hazel or brown. Legs and feet black.

The blue
Female plumage: Blue of medium shade, free from lacing. The plumage of hackles, back and shoulders in males of a darker shade.

The white
Male and female plumage: Snow white, free from straw tinge.

In both sexes: Beak white, eyes red. Legs and feet white.

In both sexes and all colours: Comb, face and wattles brilliant cherry red. Ear-lobes spotlessly white, especially near wattles.

Weights

Male   570–620 g (20–22 oz)
Female 450–510 g (16–18 oz)

Scale of points

Head (comb 20, lobes 15)	35
Tail	15
Colour	12
Type	15
Condition	15
Legs, etc.	8
	100

Serious defects

Stiltiness. Narrow chest or back. Hollow-fronted or leafy comb. Coarse bone. Tightly carried wings. Narrow feathers. Blushed lobes. Coloured feathers. White in face. In blacks, grizzled or brown flights; purple sheen or barring; light legs.

Rumpless Game Bantams

Origin: Great Britain
Classification: Light
Egg colour: Tinted

The origin of the Rumpless is unknown but like many other tail-less species it is not uncommon in the Isle of Man. Any fowl without the parson's nose or 'Caudal Appendage' can be termed rumpless but this variety has been segregated into a breed of its own.

General characteristics: male

Carriage: Bold and upright.

Type: Body short and small. Back steeply sloping. Breast full and prominent. Wings large and long. Tail completely absent, the whole of the lower back being covered with the saddle feather.

Head: Small and tapering. Beak rather large. Eyes large and bold. Comb single, small and thin, upright. Ear-lobes and wattles fine, small and thin. (Although essentially a Game variety, it is not customary to dub the males in this breed.)

Neck: Long and upright; hackle close and wiry.

Legs and feet: Legs strong and of medium length. Thighs short, round and muscular. Toes, four, long, straight and tapering.

Plumage: Hard and close.

Female

The general characteristics are similar to those of the male, allowing for the natural sexual differences.

Colour

Plumage colour is of secondary importance in this breed, and almost any recognized 'Game' colour is acceptable. Face, comb, wattles and ear-lobes should be bright red.

Weights

Male   620–740 g (22–26 oz)
Female 510–620 g (18–22 oz)

Scale of points

Type	25
Carriage and bearing	20
Head	15

(continued)

Rumpless Game Bantams

Rumpless Game
bantams
Male
Female

(**Scale of points** – *continued*)

Colour	15
Legs and feet	5
Size	15
Condition	5
	100

Disqualifications

Any sign of tail. Other than single comb. Any deformity.

Scots Dumpy

Large fowl

Origin: Great Britain
Classification: Light
Egg colour: White

This breed has been bred in Scotland for more than a hundred years, and the birds known also as Bakies, Crawlers and Creepers. Fowls having identical dumpy characters have been described as early as 1678. The breed is considered an ideal sitter and mother.

General characteristics: male

Carriage: Heavy, with a waddling gait, the extreme shortness of its legs giving the bird the appearance of 'swimming on dry land'. Shortness of leg alone should not constitute the breed's claim to notice. The large, low, heavy body, and other points of excellence must be possessed also.

Type: Body square. Back broad and flat. Breast deep. Wings of medium size and neatly carried. Tail full and flowing, the sickles well arched.

Head: Fine. Beak strong and well curved. Eyes large and clear. Comb single or rose, the former preferred. The single of medium size, upright and straight, free from side sprigs, and the back following the line of the skull, evenly serrated on top. The rose also of medium size, straight and firmly set, full of fine work or spikes, level on top, and narrowing behind to a distinct peak following the line of the skull and not sticking up. Face smooth. Ear-lobes small and close to the neck. Wattles of medium size.

Neck: Of fair length, in keeping with the size of the body, and covered with flowing hackle.

Legs and feet: Legs very short, the shanks not exceeding 3.75 cm (1½ in). Toes, four, well spread.

Scots Dumpy

Scots Dumpy large fowl Black female

Female
The general characteristics are similar to those of the male, allowing for the natural sexual differences.

Colour
Male and female plumage: There is no fixed plumage colour, but the varieties chiefly exhibited are black, cuckoo, dark, and silver grey, the last three being similar to those varieties of the Dorking. Cuckoo to be light grey with dark grey.

In both sexes: Eyes red. Comb, face, wattles and ear-lobes bright red. Beak, legs and feet white, except in the black variety where they should be black or slate; and in the cuckoo mottled.

Weights
Male   3.20 kg (7 lb)
Female 2.70 kg (6 lb)

Scale of points
Type	40
Size	20
Head	15
Condition	15
Colour	10
	100

Scots Dumpy large fowl
Cuckoo male

Serious defects

White ear-lobes. Yellow or feathered shanks or feet. Long legs. Any deformity.

Bantams

Scots Dumpy bantams have been seen occasionally at shows, and the large fowl standard should be used for them with allowance being made for size.

Scots Grey

Large fowl

Origin: Great Britain
Classification: Light
Egg colour: White

A light non-sitting breed originated in Scotland, it has not been bred extensively outside that country where, even if it is less popular today, it will doubtless be maintained by keen breeders. It has been bred there for over two hundred years.

Scots Grey

Scots Grey Male

General characteristics: male

Carriage: Erect, active and bold.

Type: Body compact, full of substance and fairly long. Back broad and flat. Breast deep, full and carried upwards. Wings moderately long and well tucked, the bow and tip covered by the neck and saddle hackles, Tail fairly long and well up (but not squirrel fashion) with full sickles.

Head: Long and fine. Beak strong and well curved. Eyes large and bright. Comb single, upright, of medium size, with well-defined serrations, the back following the line of the skull. Face of fine texture. Ear-lobes of medium size. Wattles of medium length with a well-rounded lower edge.

Neck: Finely tapered and with profuse hackle flowing on the back and shoulders.

Legs and feet: Legs long and strong. Thighs wide apart but not quite as prominent as those of Game fowl. Shanks free from feathers. Toes, four, straight and spreading, stout and strong.

Handling: Firm, and somewhat similar to the Game fowl.

Female

With the exception of the comb, either erect or falling slightly over, the general characteristics are similar to those of the male, allowing for the natural sexual differences.

Colour

Male plumage: Cuckoo feathered. Ground colour of body, thighs and wing feathers steel grey. The barring is black with a metallic lustre, that of the body, thighs and wing feathers straight across, but that of the neck hackle, saddle, and tail slightly angled or V-shaped. The alternating bands of black are of equal width and proportioned to the size of the feather. The bird should 'read' throughout, i.e. the shade should be the same from head to tail. The plumage should be free from red, black, white, or yellow feathers, and the hackle, saddle, and tail should be distinctly and evenly barred, while the markings all over should be rather small, even, and sharply defined.

Female plumage: Similar to that of the male, except that the markings are not as small, and produce an appearance somewhat resembling a shepherd's tartan.

In both sexes: Beak white or white streaked with black. Eyes amber. Comb, face, wattles and ear-lobes bright red. Legs and feet white or white mottled with black, but not sooty.

Weights

Male 3.20 kg (7 lb)
Female 2.25 kg (5 lb)

Scale of points

Colour and markings	30
Size	10
Type	30
Head	10
Condition	10
Legs and feet	10
	100

Serious defects

Any bodily deformity. Any characteristic of any other breed not applicable to the Scots Grey.

Bantams

Scots Grey bantams follow the large fowl standard.

Weights

Male 620–680 g (22–24 oz)
Female 510–570 g (18–20 oz)

Sebright Bantams

Origin: Great Britain
Classification: True bantam
Egg colour: White or cream

This breed is a genuine bantam and one of the oldest British varieties. It has no counterpart in large breeds, but has played a part in the production of other laced fowl, notably Wyandottes. There are two colours, gold and silver.

General characteristics: male

Carriage: Strutting and tremulous, on tip-toe, somewhat resembling a fantail pigeon.

Type: Body compact, with broad and prominent breast. Back very short. Wings large and carried low. Tail square, well spread and carried high. Sebright males are hen feathered, without curved sickles or pointed neck and saddle hackles.

Head: Small. Beak short and slightly curved. Comb rose, square fronted, firmly and evenly set on, top covered with fine points, free from hollows, narrowing behind to a distinct spike or leader, turned slightly upwards. Eyes full. Face smooth. Ear-lobes flat, and unfolded. Wattles well rounded.

Neck: Tapering, arched and carried well back.

Legs and feet: Legs short and well apart. Shanks slender and free from feathers. Toes, four, straight and well spread.

Plumage: Short and tight, feathers not too wide but never pointed. (Almond-shaped feather is desired.)

Female

The general characteristics are similar to those of the male, allowing for the natural sexual differences. Her neck is upright.

Colour

The gold
Male and female plumage: Uniform golden bay with glossy green-black lacing and dark grey undercolour; each feather evenly and sharply laced all round its edge with a narrow margin of black. Shaftiness is undesirable.

The silver
Male and female plumage: Similarly marked on pure, clear silver white ground colour.

In both sexes and colours: Beak dark horn in golds; dark blue or horn in silvers. Eyes black, or as dark as possible. Comb, face, wattles and ear-lobes dark purple or dull red (mulberry). Legs and feet slate blue.

Sebright Bantams

Sebright bantams
Gold male
Silver female

Shamo

Although in males the purple or mulberry face is seldom obtainable, the eye should be dark and surrounded with a dark cere.

Weights

Male 620 g (22 oz)
Female 510 g (18 oz)

Scale of points

Lacing	25
Comb	5
Face and lobes	10
Ground colour	15
Tail	10
Type	20
Weight	5
Condition	10
	100

Note: There is at present a decided move to improve type and to discourage the prevailing whip tails and narrow build, particularly in females.

Serious defects

Single comb. Sickle feathers or pointed hackles on the male. Feathers on shanks. Legs other than slate blue. Other than four toes. Any deformity.

Shamo

Origin: Asia
Classification: Heavy
Egg colour: Tinted

The Shamo is a variation on the Malay with slightly shorter legs and little or no curve to the back. It is of Japanese origin and arrived in this country in the early 1970s. Black-red is the most popular colour.

General characteristics: male

Carriage: Very upright.

Type: General appearance big and powerful. Large firm body. Broad and square. Back sloping. Wings large and strong carried close. Tail carried horizontally or a little below.

Head: Deep and broad without wattles or lobes. Beak strong and broad. Eyes deep set under overhanging brows. Comb walnut or triple.

Shamo large fowl
Male

Neck: Long and thick.

Legs and feet: Thighs long. Shanks medium length, strong boned with a slight bend at the hock. Toes, four, long and well spread.

Plumage: Short and hard.

Handling: Firm fleshed and muscular.

Female

The general characteristics are similar to those of the male, allowing for the natural sexual differences.

Colour

Male and female plumage: Any game colour.

In both sexes and all colours: Beak, legs and feet yellow. Comb, face, throat and ear-lobes brilliant red. Eyes very light to orange.

Sicilian Buttercup

Weights

Male 5.00–5.45 kg (11–12 lb) min.
Female 2.95 kg (6½ lb) min.

Scale of points

Type and carriage	40
Head	20
Feathering and condition	20
Colour	10
Legs	10
	100

Serious defects

Lack of size. Poor carriage. Wry tail.

Sicilian Buttercup

Origin: Continental Europe
Classification: Light
Egg colour: White

The Sicilian Buttercup was first imported into this country in 1912 by Mrs. Colbeck of Yorkshire. There is never any mistaking this breed because of its distinctive saucer-shaped cup comb. The Buttercup was plentiful in the 1960s when many were used for laying trials because of their ability to produce many eggs from little food.

General characteristics: male

Carriage: Upright, bold and active.

Type: Body moderately long and deep; broad shoulders and narrow saddle. Full round breast. Broad back sloping downwards to the saddle, which rises in a slight concave outline to the base of the tail. Long wings, closely tucked. Fairly large tail with long main feathers, carried at an angle of 45°, and fitted with well-curved sickles and abundant coverts.

Head: Skull fairly long and deep. Beak of medium length. Eyes full and keen. Comb beginning at the base of the beak with a single leader and joined to a cup-shaped crown, set firmly on the centre of the skull and surmounted with well-defined and regular points, of medium size and fine texture, and free from decided spikes in the cavity or centre. Ear-lobes almond-shaped, flat, smooth, and close fitting. Wattles thin and well rounded.

Sicilian Buttercup

Sicilian Buttercup large fowl Male

Neck: Rather long, with hackle flowing well over the shoulders.

Legs and feet: Of moderate size and length. Thighs well apart. Shanks slender and free of feathers. Toes, four, straight and spreading.

Female

With the exception of the comb (smaller and lower in proportion) the general characteristics are similar to those of the male, allowing for the natural sexual differences.

Colour

The brown
Male plumage: Neck hackle rich orange-red striped with black, crimson red in front below wattles. Back and shoulder coverts deep crimson red or maroon. Saddle rich orange-red with or without a few black stripes. Wing bow deep crimson red or maroon; coverts steel blue with green reflexions

Sicilian Buttercup

forming the bar; primaries brown, secondaries deep bay on outer web, black on inner. Breast and underparts glossy black quite free from brown splashes. Tail green-black, coverts black edged with brown.

Female plumage: Hackle rich golden yellow, broadly striped with black. Breast salmon red running into maroon near the wattles and ash grey at thighs. Body and wings rich brown, very finely and closely pencilled with black, free from any red tinge. Tail black pencilled with brown.

(*Note:* The undercolour of both sexes should be slate blue or mouse.)

The golden

Male plumage: Neck hackle, back, saddle, shoulders and wing bows bright lustrous orange-red. Cape (at base of neck) dark buff marked with distinct black spangles and covered by hackle. Wing bar and bay an even shade of red-bay; primaries black, lower web edged with bay; secondaries red-bay on outer web, black on inner. Breast red-bay. Body light bay. Fluff rich bay shading to light bay on stern, and some feathers on the body fluff with distinct black spangles. Tail black, sickles and coverts green-black, the former showing red-bay at their base and the coverts edged with that colour.

Female plumage: Hackle lustrous golden buff. Breast and thighs light golden buff, plain from throat to middle of breast, elsewhere with black spangles. Tail dull black, except the two highest feathers mottled with buff. Wing bow and bar golden buff with parallel rows of elongated black spangles, each spangle extending slightly diagonally across the web; quill and edge of feathers golden buff; primaries black edged with buff; secondaries golden buff regularly barred with black on outer web, black on inner. Back golden buff regularly spangled with black (the same pattern as the wing bow) and extending over the entire surface, including the saddle and the tail coverts.

The silver

Male and female plumage: Except that the ground is silver white (free from yellow or straw tinge) instead of red or bay, similar to the golden.

The golden duckwing

Male plumage: Neck and saddle hackles white with cream tinge. Back and saddle, maroon-orange shading to cream. Wing butts and bars black; diamond white; bows bright orange to cream. Remainder black.

Female plumage: Hackle white, fairly striped with black or grey. Tail black, except top feathers slightly pencilled with grey. Remainder, slate grey, finely pencilled with darker grey, or black.

The white

Male and female plumage: Pure white throughout.

In both sexes and all colours: Beak dark horn lightly shaded with yellow. Eyes red-bay. Comb, face, wattles and ear-lobes bright red (more than one-third white in lobes a serious defect). Legs and feet willow green.

Weights

Cock 2.95 kg (6½ lb); cockerel 2.50 g (5½ lb)
Hen 2.50 kg (5½ lb); pullet 1.80 kg (4 lb)

Scale of points

The brown

Head (comb 15, wattles and lobes 6, eye 4, beak 4)	29
Colour	25
Neck hackle	10
Type	10
Size	10
Condition	10
Legs and feet	6
	100

Other varieties

Colour (body and fluff, and back 6 each, wings 5, head, wattles and lobes, neck, breast, and tail 4 each, legs and feet 3, eyes and beak 2 each)	44
Type (tail and breast 4 each, neck, wings, back, and legs and feet 3 each, head, eyes, beak, wattles and lobes, and body and fluff 2 each)	30
Comb	14
Symmetry	4
Size	4
Condition	4
	100

Serious defects

Spikes more than 2.5 cm (1 in) long in cup (or cavity of comb) of male, or any indication of spike in cup (or cavity of comb) of female. Solid white ear-lobes. Shanks other than green. Feathers or stubs on shanks or toes. Any deformity. In golden, solid white in any part of the plumage (except in undercolour) or black striping in the male's hackles; in browns, white in male's tail.

Silkie

Large fowl

Origin: Asia
Classification: Light
Egg colour: Tinted to cream

Silkie fowls have been mentioned by authorities for several hundred years, although some think they originated in India, while others favour China

Silkie

Silkie large fowl
White male
Black female

and Japan. It seems likely that such oddities came from Japan. Despite such small weights of 0.90–1.35 kg (2–3 lb) the Silkie is not regarded as a bantam in this country but as a light breed, and as such it must be exhibited. Its persistent broodiness is a breed characteristic, and either pure or crossed the breed provides reliable broodies for the eggs of large fowl or bantams.

General characteristics: male

Carriage: Stylish, compact and lively.

Type: Body broad and stout looking. Back short, saddle silky and rising to the tail, stern broad and abundantly covered with fine fluff, saddle hackles soft, abundant, and flowing. Breast broad and full; shoulders stout, square, and fairly covered with neck hackle. Wings soft and fluffy at the shoulders, the ends of the flights ragged and 'osprey plumaged' (i.e. some strands of the flight hanging loosely downward). Tail short and very ragged at the end of the harder feathers of the tail proper. It should not be flowing, but a short round curve.

Head: Short and neat, with good crest, soft and full, as upright as the comb will permit, and having half a dozen to a dozen soft, silky feathers streaming gracefully backwards from lower and back part of crest, to a length of about 3.75 cm (1½ in). The crest proper should not show any hardness of feathers. Beak short and stout at base. Eyes brilliant and not too prominent. Comb almost circular in shape, preferably broader than long, with a number of small prominences over it; preferably having a slight indentation or furrow transversely across the middle. Face smooth. Ear-lobes more oval than round. Wattles concave, nearly semi-circular, not long or pendent.

Neck: Short or medium length, broad and full at base with the hackle abundant and flowing.

Legs and feet: Free from scaliness. Thighs wide apart and legs short. No hard feathers on the hocks but a profusion of soft silky plumage on them is admissible. The feathers on the legs should be moderate in quantity. Toes, five, the fourth and fifth diverging from one another. The middle and outer toes feathered, but these feathers should not be too hard. Thighs covered with abundant fluff.

Plumage: Very silky and fluffy with a profusion of hair-like feathers.

Female

Saddle broad and well cushioned with the silkiest of plumage which should nearly smother the small tail, the ragged ends alone protruding, and inclined to be 'Cochiny' in appearance. The legs are particularly short in the female, in which the underfluff and thigh fluff should almost meet the ground. The head crest is short and neat, like a powder puff, with no hard feathers, nor should the eye be hidden by the crest, which should stand up and out, not split by the comb. Ear-lobes small and roundish. Wattles

Silkie

either absent or small and oval in shape. Other general characteristics are similar to those of the male, allowing for the natural sexual differences. Comb small.

Colour

The black
Male and female plumage: Black all over with a green sheen in the males; colour in hackle is permissible.

The blue
Male and female plumage: An even shade of blue from head to tail; a self colour, and neither laced nor barred.

The gold
Male and female plumage: An even shade of golden buff, avoiding pale lemon colour on the one hand and brownish orange on the other. Clear colour throughout to be preferred but some darker feathers permissible in tails of both sexes.

The white
Male and female plumage: Snow white.

In both sexes and all colours: Beak slaty blue. Eyes black. Comb, face and wattles mulberry. Ear-lobes turquoise blue or mulberry, the former preferred. Legs and feet lead. Nails blue-white. Skin mulberry.

Weights

Male   1.35 kg (3 lb)
Female 0.90 kg (2 lb)

(There is no objection to size in either sex if the other points are good.)

Scale of points

Type	20
Head	30
Legs	10
Colour	10
Plumage	30
	100

Serious defects

Hard feathers. Vulture hocks. Green beak or green tip to beak. Horns protruding from comb. Ruddy comb or face. Eye other than black. Incorrect colour in plumage or skin. Plumage not silky. Want of crest; 'Polish' or 'split' crest; the crest should not hang over the eyes. Green soles to feet.

Disqualifications

Single comb. Green legs. Four toes. Featherless legs and feet.

Bantams

Silkies are not standardized in bantam form.

Spanish

Large fowl

Origin: Mediterranean
Classification: Light
Egg colour: White

The white-faced black Spanish is one of our oldest breeds, and was widely kept and admired long before the advent of poultry shows in the latter half of the nineteenth century. Of striking appearance, with its extensive white face, surrounding eyes and ears and extending lower than the wattles, the Spanish was also a good layer of large white eggs. The emergence of the red-faced Minorca pushed the Spanish into the background, but they still appear occasionally at shows.

General characteristics: male

Carriage: Upright, with proud action.

Head: Skull long, broad and deep. Beak long and stout. Eyes full and wide open. Comb single, somewhat small, erect and straight, firm at the base, rather thin at the edge, fitting closely on the neck at the back, of very smooth texture, and free from wrinkles, rising well over the eyes but not so as to interfere with the sight, and joining the ear-lobes and wattles. Ear-lobes deep and broad, well rounded at the bottom, extending well below the wattles, meeting in front and going well back on each side of the neck, of fine texture and free from folds or creases. Wattles very long, thin, and pendulous.

Neck: Long and fine, with abundant hackle flowing well over the shoulders.

Body: Rather long, fairly broad in front, and tapering to the rear. Breast full at the neck and gradually decreasing towards the thighs. Back slanting downwards to the tail, short wings carried closely. Full tail, not carried too high, and with the sickles large and well curved.

Legs and feet: Rather long and slim. Shanks free of feathers. Toes, four, slender and straight.

Plumage: Short and close.

Female

With the exception of the comb (which falls gracefully over either side of the face) the general characteristics are similar to those of the male, allowing for the natural sexual differences.

Colour

Male and female plumage: Black with a beetle green sheen, and free of purple bars.

Spanish

Spanish large fowl Male

In both sexes: Beak dark horn. Eyes black. Comb and wattles bright red. Face and ear-lobes white. Legs and feet pale slate.

Weights

Male   3.20 kg (7 lb)
Female 2.70 kg (6 lb)

Scale of points

Face and lobes	35
Comb and wattles	15
Type	15
Size	15
Colour	10
Condition	10
	100

Serious defects

Blue, pink or red in face or lobes. Coarse 'cauliflower' face or lobes. Male's comb not erect, side sprigs on comb. Lobes pointed at the bottom. Black or dark legs or feet. Any deformity.

Bantams

Should Spanish bantams be bred they should follow the large fowl standard.

Sultan

Large fowl

Origin: Turkey
Classification: Light
Egg colour: White

Sultan fowls were imported in 1854 from Constantinople. Though they never became widespread, Sultans have remained in existence and are occasionally seen at shows. In crest, beard and plumage they are similar to white Polands. However, they differ in some important respects: vulture hocks, five toes and short back. The Sultan is a breed of considerable character, tame, yet sprightly.

General characteristics: male

General shape and carriage: Deep, but neat and compact, and very sprightly.

Type: Body rather long and very deep. Breast deep and prominent. Back short and straight. Wings large, long, and carried low. Tail long, broad, and carried open. Sickles very long and fine. Hangers numerous, long and fine. Coverts abundant and lengthy.

Head: Head medium size. Beak short and curved. Eye bright. Comb very small, consisting of two spikes only, almost hidden by crest. Face covered with thick muffling. Nostrils horny and large, rising above the curved line of the beak. Crest large, globular, and compact. Ear-lobes small and round. Beard very full, joining with the whiskers. Wattles very small, to be hardly perceptible.

Neck: Moderately short, slightly arched and carried well back.

Legs and feet: Thighs short, furnished with heavy vulture hocks to cover the joints. Shanks short, and well covered inside and out with feathers. Toes, five in number and of moderate length, completely covered with feather.

Sultan

Sultan large fowl Male

Sultan large fowl Female

Plumage: Long, very abundant and fairly soft.

Female

With the exception of the comb, which is smaller and barely visible, the general characteristics are similar to those of the male, allowing for the natural sexual differences.

Colour

Male and female plumage: Snow white throughout.

In both sexes: Beak pale blue or white. Eye red. Comb bright red. Face red. Ear-lobes and wattles bright red. Shanks and toes white or pale blue.

Weights

Male   2.70 kg (6 lb) max.
Female 2.00 kg (4½ lb)

Scale of points

Head and crest	15
Beard and muffling	15
Comb	5
Type and symmetry	12
Colour	15
Leg and foot feathering	15
Size	8
Condition	15
	100

Serious defects

Any deformity. Coloured plumage. Toes other than five in number.

Bantams

Sultan bantams are not standardized at present.

Sumatra Game

Large fowl

Origin: Asia
Classification: Light
Egg colour: White

Sumatra Game were admitted to the American Standard in 1883, under the name of Black Sumatra. They reached Britain later and have been bred here in small numbers ever since. A long flowing tail, carried horizontally, and a pheasant-like carriage are two of their main characteristics. Sumatra Game are prolific layers, and excellent sitters.

Sumatra Game

Sumatra Game large fowl
Male

General characteristics: male

Carriage: Straight and upright in front, pheasant-like, giving a proud and stately appearance.

Type: Body rather long, very firm and muscular, broad, full and rounded breast. Back of medium length, broad at shoulders, very slightly tapering to tail. Saddle hackle very long and flowing. Stern narrower than shoulders, but firm and compact. Strong, long, and large wings, carried with fronts lightly raised, the feathers folded very closely together, not carried drooping or over the back. Long drooping tail with a large quantity of sickles and coverts, which should rise slightly above the stern and then fall streaming behind, nearly to the ground. Sickle and covert feathers not too broad.

Head: Skull small, fine, and somewhat rounded. Beak strong, of medium length, slightly curved. Eyes large and very bright, with a quick and fearless expression. Comb pea, low in front, fitting closely, the smaller the better. Face smooth and of fine texture. Ear-lobes and wattles as small as possible and fitting very closely.

Neck: Rather long, and covered with very long and flowing hackle.

Sumatra Game

Legs and feet: Of strictly medium length, thick and strong. Thighs muscular, set well apart. Shanks straight and strong, set well apart, with smooth, even scales, not flat or thin. (*Note:* There is no objection to two or more spurs on each leg, it being a peculiarity of the breed for this to occur.) Feet broad and flat. Toes, four, long, straight, spread well apart, with strong nails, the back toe standing well backward and flat on the ground.

Plumage: Very full and flowing, but not too soft or fluffy.

Female

The general characteristics are similar to those of the male, allowing for the natural sexual differences.

Colour

Male and female plumage: Very rich beetle green (green-black) with as much sheen as possible.

In both sexes: Beak dark olive or black (olive preferred). Eyes very dark red, dark brown, or black (dark red preferred). Face, comb, ear-lobes and wattles black or 'gypsy' faced or very dark red (gypsy face preferred). Legs and feet dark olive or black (olive preferred).

Weights

Male 2.25 kg (5 lb)
Female 1.80 kg (4 lb)

Scale of points

Type	20
Head (beak 5, eyes 5, other points 10)	20
Colour	15
Feather, quantity of	15
Condition	15
Legs and feet	10
Neck	5
	100

Serious defects

Single or rose comb, dubbing. Other than four toes. Any deformity.

Bantams

The standard for large fowl should be used for Sumatra Game bantams.

Sussex

Large fowl

Origin: Great Britain
Classification: Heavy
Egg colour: Tinted

This is a very old breed, for although we do not find it included in the first Book of Standards of 1865, yet at the first poultry show of 1845 the classification included Old Sussex or Kent fowls, Surrey fowls and Dorkings. The oldest variety of the Sussex is the speckled. Brahma, Cochin, and silver grey Dorking were used in the make-up of the light. The earlier reds had black breasts, until the red and brown became separate varieties. Old English Game has figured in the make-up of some strains of browns. Buffs appeared about 1920, clearly obtained by sex-linkage within the breed. Whites came a few years later, as sports from lights. Silvers are the latest variety. The light is the most widely kept in this country today among standard as well as commercial breeders. It is one of our most popular breeds for producing table birds. At the time when sex-linkage held considerable popularity the light Sussex was one of the most popular breeds of the day, the females being in considerable demand for mating to gold males. At an even earlier stage the Sussex breed formed the mainstay of the table poultry market in and around the Heathfield area. The Sussex Breed Club was formed as far back as in 1903 and is now one of the oldest breed clubs in Britain.

General characteristics: male

Carriage: Graceful, showing length of back, vigorous and well balanced.

Type: Back broad and flat. Breast broad and square, carried well forward, with long, straight and deep breast bone. Shoulders wide. Wings carried close to the body. Skin clear and of fine texture. Tail moderate size, carried at an angle of 45°.

Head: Of medium size and fine quality. Beak short and curved. Eyes prominent, full and bright. Comb single, of medium size, evenly serrated and erect, and fitting close to the head. Face smooth and of good texture. Ear-lobes and wattles of medium size and fine texture.

Neck: Gracefully curved with fairly full hackle.

Legs and feet: Thighs short and stout. Shanks short and strong, and rather wide apart, free from feather, with close-fitting scales. Toes, four, straight and well spread.

Plumage: Close and free from any unnecessary fluff.

Female

The general characteristics are similar to those of the male, allowing for the usual sexual differences.

Sussex

Sussex large fowl
Speckled male
Speckled female

Sussex

Sussex large fowl
Silver male
Brown female

Sussex

Sussex large fowl
Light male
Light female

Sussex

Colour

The brown
Male plumage: Head and neck hackles rich dark mahogany striped with black. Saddle hackle same as neck hackle. Back and wing bow rich dark mahogany. Wing coverts forming the bar blue-black; secondaries and flights black, edged with brown. Breast, tail and thighs black.

Female plumage: Head and neck hackles brown striped with black. Back and wings dark brown, finely peppered with black. Breast and underbody clear pale wheaten brown. Flights black, edged with brown. Tail black.

The buff
Male and female plumage: Body rich even golden buff. Head and neck hackles buff, sharply striped with green-black. Wings buff, with black in the flights. Tail and coverts greeny black. Dark in undercolour, not penalized at present, but buff is desirable.

The light
Male and female plumage: Head and neck hackles white, striped with black, the black centre of each feather to be entirely surrounded by a white margin. Wings white, with black in flights. Tail and coverts black. Remainder pure white throughout.

The red
Male and female plumage: Head and neck hackles rich dark red, striped with black. Body and wing bow rich dark red, one uniform shade throughout free from pepperiness. Wings rich dark red with black in the flights. Tail black, coverts rich dark red. Undercolour slate.

The speckled
Male plumage: Head and neck hackles rich dark mahogany, striped with black and tipped with white. Wing bow speckled, primaries white, brown and black. Saddle hackle similar to neck hackle. Tail, main feathers black and white, sickles black with white tips. Remainder rich dark mahogany, each feather tipped with a small white spot, a narrow glossy black bar dividing the white from the remainder of the feather. Undercolour slate and red with a minimum of white.

Female plumage: Head, neck and body ground colour rich dark mahogany, each feather tipped with a small white spot, a narrow glossy black bar dividing the white from the remainder of feathers, the mahogany part of feather free from pepperiness, neither of the colours to run into each other, and to show the three colours distinctly; undercolour as for male. Tail black and brown with white tip. Flights black, brown and white.

The silver
Male plumage: Head, neck and saddle hackles white striped with black, the black centre of each feather to be entirely surrounded by a white margin. Wing bow and back silvery white; coverts forming bar black; flights and secondaries black tinged with grey. Breast black with white shafts, and silver lacing round feathers. Thighs dark grey showing faint lacing. Tail black. Undercolour grey-black shading to white at skin.

Female plumage: Head and neck hackles as in the male. Back and wing bow greyish black, each feather showing white shaft with fine silver lacing

Sussex

Sussex bantams
Silver male
Silver female

Sussex

surrounding it; flights and secondaries greyish black. Tail black. Breast and thighs lighter shade of greyish black with white shafts and silver lacing to correspond with the top colour. Undercolour as in the male.

The white
Male and female plumage: Pure white throughout and to the skin.

In both sexes and all colours: Beak white or horn generally, dark or horn with the brown and dark shading to white with the silver. Eyes: brown – brown or red; buff, red and speckled – red; light, silver and white – orange. Face, comb ear-lobes and wattles red. Shanks and feet white. Flesh and skin white.

Weights

Male   4.10 kg (9 lb) min.
Female 3.20 kg (7 lb) min.

Scale of points

Type and flatness of back	25
Size	20
Legs and feet	15
Colour	20
Condition	10
Head and comb	10
	100

Serious defects (for which birds should be passed)

Other than four toes. Wry tail or any other deformity. Feather on shanks and toes. Rose comb.

Bantams

Sussex bantams conform to the large fowl standard.

Weights

Male   1130 g (40 oz) max.
Female 790 g (28 oz) max.

Scale of points

Type, size and weight	35
Hackle, tail and wing	20
Body colour	15
Head and eye	15
Feet and legs	15
	100

Transylvanian Naked Neck

Large fowl

Origin: Europe
Classification: Heavy
Egg colour: Tinted

The Naked Neck, as this breed is more commonly called, have also been called 'Churkeys'. Strains in this country at the moment come from Malta where for many years they found fame in their ability to live and lay well on the most frugal diet. An interesting behavioural trait is that they dig deep into rubbish piles, dung heaps and the like for food.

General characteristics: male

Carriage: Alert, upright and bold.

Type: Body large, deep and compact, well balanced and symmetrical. Back broad and of medium length, saddle hackle long and abundant. Breast broad and well rounded. Wings of medium size, carried well up. Tail medium size carried at an angle of 45°, sickles large and well curved.

Head: Medium size. Beak short, stout and slightly curved. Eyes large, bright and prominent. Comb single, medium in size, straight and erect, with well-formed spikes. Face smooth. Ear-lobes and wattles of medium size, fine in texture and smooth: the head to carry an oval cap of feathers surrounding the base of the comb, even in shape.

Neck: Of medium length, slightly curved, completely without feathering, stubs or fluff: the skin of the neck to be smooth and fine in texture, free from wrinkles or roughness. (A small tassel of feathers at the bottom of the neck above the breast feathers is permitted, but not desirable.)

Legs and feet: Legs of medium length, strong and stout. Shanks free of feathers. Toes, four, strong, straight and well spread.

Female

The general characteristics are similar to those of the male, allowing for the natural sexual differences.

Colour

The black
Male and female plumage: Dense black with a rich green sheen.

The white
Male and female plumage: Snow white throughout.

The cuckoo
Male and female plumage: Ground colour silvery white: each feather to be evenly and distinctly barred with black with a green sheen.

Transylvanian Naked Neck

Transylvanian Naked Neck Male

Transylvanian Naked Neck Female

The buff
Male and female plumage: Colour as in buff Rocks.

The red
Male and female plumage: Colour as in Rhode Island Reds.

The blue
Male and female plumage: Colour as in Andalusians.

In both sexes and all colours: Eyes orange. Face, comb, ear-lobes, wattles and neck bright red. Legs, feet and beak yellow or horn in the black and cuckoo, yellow or white in the white.

Weights

Male   3.60–4.55 kg (8–10 lb)
Female 2.70–3.60 kg (6–8 lb)

Scale of points

Type and carriage	20
Head and neck	35
Legs and feet	10
Colour and markings	15
Size	10
Condition	10
	100

Serious defects

Any noticeable feather, fluff or stubs on the neck. Absence of cap of feathers on the head. Feathered legs. Other than four toes. Any deformity.

Bantams

Bantam Naked Necks are to be replicas of their large fowl counterparts.

Weights

Male   910 g (32 oz)
Female 680 g (24 oz)

Tuzo Bantams

Origin: Japan
Classification: True bantam
Egg colour: Tinted

The Tuzo is a true hard feather bantam from Japan. It has been in this country since the early 1970s and is still in a few hands. It is not unlike a bantam Asil.

Tuzo Bantams

General characteristics: male

Carriage: Upright.

Type: Similar to a small Asil. Body broad at front with prominent shoulders. Wings short. Tail carried horizontally or a little below. Feathers short and hard.

Head: Broad and rounded with a short, slightly hooked beak. Well developed brows and protruding cheeks. Comb small triple or occasionally walnut. Wattles and lobes (if any) insignificant.

Neck: Strong and slightly curved.

Legs and feet: Thighs strong with slight bend at hock. Shanks medium strong and straight. Toes, four, fine and straight.

Female

The general characteristics are similar to those of the male, allowing for the natural sexual differences. The comb, if visible, should be small and insignificant.

Colour

Male and female plumage: Any Game colour.

In both sexes and all colours: Eyes light yellow to orange. Comb, face, wattles and lobes bright red.

Weights

Male   1020–1250 g (36–44 oz)
Female 790–1020 g (28–36 oz)

Scale of points

Type and carriage	40
Plumage quality	10
Colour	10
Head	20
Legs and feet	10
Condition	10
	100

Serious defects

Any deformity. Comb any other than standard.

Vorwerk

Large fowl

Origin: Germany
Classification: Light
Egg colour: Cream to tinted

Originated in Hamburg by Oskar Vorwerk in 1900, the breed was first shown at Hanover in 1912 and standardized in 1913. The aim was to provide a middle-weight economical utility fowl, good natured, lively but not timid. A point worthy of note is the compatibility of males amongst themselves. These fowls were found to be particularly suitable for small-holdings and farmyards as they are excellent foragers, small eaters and quick maturing.

General characteristics: male

Carriage: Very powerful, compact utility shape, carriage low rather than high, not too much bone, markings the same in both sexes, lively but not timid.

Type: Body of considerable size, as broad and deep as possible like a rounded rectangle. Back broad, slightly sloping with a full saddle. Breast broad, deep and well rounded. Wings closely carried. Tail moderately tight, held at a lowish angle with well-rounded sickles of moderate length.

Head: Medium sized and moderately broad. Face covered with small feathers. Comb single, of medium size at the most, with four to six serrations. Wattles of medium length, well rounded. Lobes of barely average size and white. Beak greyish blue to horn. Eyes alert, orange to orange-red.

Neck: Of moderate length with full hackle and carried fairly upright, proudly.

Legs and feet: Moderate length with fine bone. Toes, four, small close fitting scales. Thighs fleshy and tightly feathered.

Plumage: Close fitting, glossy, velvety hackle.

Handling: Firm as befits an active forager.

Female

General characteristics are similar to those of the male allowing for the natural sexual differences. Back to be broad with almost no cushion. The latter part of the small comb may bend slightly to one side.

Colour

Male plumage: Head, hackle and tail should be velvety black. Body deep buff, undercolour grey. Wing secondaries buff, primaries dark grey to black. Saddle buff with light striping. Legs slate.

Vorwerk

Vorwerk large fowl Male

Vorwerk large fowl Female

Female plumage: Hackle black with slight buff lacing permitted at the back of the head. Body and secondary wing flights buff. Primaries greyish black and buff mixed. Visible parts of the main tail black with the tail furnishings partly laced with buff. Undercolour grey.

In both sexes: Beak greyish blue to horn. Eyes orange or orange-red. Comb, face and wattles red. Lobes white. Legs and feet slate.

Weights

Male   2.50–3.20 kg (5½–7 lb)
Female 2.00–2.50 kg (4½–5½ lb)

Scale of points

Type/utility quality	25
Head	10
Colour	25
Legs and feet	10
Size	15
Condition	15
	100

Serious defects

Body too narrow or too light. Carriage too high. Coarse bone. High tail. Lobes too red. Pale legs. In males, hackle unduly buff or grey, saddle nearly black. In females, lack of black in neck or tail and undue spangling in body feathers.

Bantams

The same standard as in large fowl applies to bantams, the only difference being in the weights.

Weights

Male   910 g (32 oz)
Female 680 g (24 oz)

Welsummer

Large fowl

Origin: Holland
Classification: Light
Egg colour: Brown to deep brown

Named after the village of Welsum, this Dutch breed has in its make-up such breeds as the partridge Cochin, partridge Wyandotte and partridge

Welsummer

Leghorn, and still later the Barnevelder and the Rhode Island Red. In 1928, stock was imported into this country from Holland, in particular for its large brown egg, which remains its special feature, some products being mottled with brown spots. It has distinctive markings and colour, and comes into the light-breed category, although it has good body-size. It enters the medium class in the country of its origin. Judges and breeders work to a standard that values indications of productiveness, so that laying merits can be combined with beauty.

General characteristics: male

Carriage: Upright, alert and active.

Type: Body well built on good constitutional lines. Back broad and long. Breast full, well rounded and broad. Wings moderately long, carried closely to the sides. Tail fairly large and full, carried high, but not squirrel. Abdomen long, deep and wide.

Head: Symmetrical, well balanced, of fine quality without coarseness, excesses or exaggeration. Skull refined, especially at back. Beak strong, short and deep. Eyes keen in expression, bold, full, highly placed in skull and standing out prominently when viewed from front or back; pupils large and free from defective shape. Comb single, of medium size, firm, upright, free from any twists or excess around nostrils, clear of nostrils, and of fine, silky texture, five to seven broad and even serrations, the back following closely but not touching the line of the skull and neck. Face smooth, open and of silky texture, free from wrinkles or surfeit of flesh and without overhanging eyebrows. Ear-lobes small and almond-shaped. Wattles of medium size, fine and silky texture and close together.

Neck: Fairly long, slender at top but finishing with abundant hackle.

Legs and feet: Thighs to show clear of body without loss of breast. Shanks of medium length, medium bone and well set apart, free from feathers and with soft, pliable sinews, free from coarseness. Toes, four, long, straight and well spread out, back toe to follow in straight line, free from feathers between toes.

Plumage: Tight, silky and waxy, free from excess or coarseness, silky at abdomen and free from bagginess at thighs.

Handling: Compact, firm and neat bone throughout.

Female

The general characteristics are similar to those of the male, allowing for the natural sexual differences. Handling: Pelvic bones fine and pliable; abdomen pliable; flesh and skin of fine texture and free from coarseness; plumage sleek; abdomen capacious, but well supported by long breast bone and not drooping; general handling of a fit, keen and active layer.

Welsummer

Welsummer large fowl
Male
Female

Welsummer

Colour

Male plumage: Head and neck rich golden brown. Hackles rich golden brown as uniform as possible, free from black striping, yet underparts (out of sight) may show a little striping at present. Back, shoulder coverts and wing bow bright red-brown. Wing coverts black with green sheen forming a broad bar across (a little brown peppering at present permissible); primaries (out of sight when wing is closed), inner web black, outer web brown; secondaries, outer web brown, inner web black with brown peppering. Tail (main) black with a beetle green sheen; coverts, upper black, lower black edged with brown. Breast black with red mottling. Abdominal and thigh fluff black and red mottled.

Female plumage: Head golden brown. Hackle golden brown or copper, the lower feathers with black striping and golden shaft. Breast rich chestnut red going well down to the lower parts. Back and wing bow reddish brown, each feather stippled or peppered with black specks (i.e. partridge marking), shaft of feather showing lighter and very distinct. Wing bar chestnut brown; primaries, inner web black, outer brown, secondaries, outer web brown, coarsely stippled with black; inner web black, slightly peppered with brown. Abdomen and thighs brown with grey shading. Tail black, outer feathers pencilled with brown.

The silver duckwing
Male plumage: Head, neck and hackles white. Breast black with white mottling. Back shoulder coverts and wing bow white. Wing primaries, flight feathers (out of sight as wing is closed), inner web black, outer web white; secondaries, outer web white, inner web black, with white peppering, coverts black with green sheen forming a broad bar across primaries. Tail, main black with beetle green sheen; coverts, upper black, lower black, edged with white. Abdominal and thigh fluff black with white mottling.
Female plumage: Head and skull silvery white. Hackle silvery white and lower feathers with black striping, and white shaft. Breast salmon red or robin red. Back and wing bow silvery grey, each feather stippled or peppered with black specks (i.e. partridge marking), shaft of feather showing light and very distinct. Wing bar silvery grey; primaries, inner web black, outer web white; secondaries, outer web white, coarsely stippled with black, inner web black slightly peppered with white. Abdomen and thighs silvery grey. Tail black, outer feathers pencilled with white.

In both sexes and colours: Beak yellow or horn. Eyes red. Comb, face, ear-lobes and wattles bright red. Legs and feet yellow. Undercolour dark slate grey.

Weights

Cock 3.20 kg (7 lb); cockerel 2.70 kg (6 lb)
Hen 2.70 kg (6 lb); pullet 2.00–2.25 (4½–5 lb)

Scale of points

General type	20
Handling, size, and indications of productiveness	30
Head	10
Legs and feet	10
Colour	20
Condition	10
	100

Serious defects

Comb other than single or with side sprigs. White in lobe. Feather on legs, hocks or between toes. Other than four toes. Striping in neck hackle or saddle of male. Absolutely black or whole red breast in the male. Salmon breast in the female. Legs other than yellow. Badly crooked or duck toes. Any body deformity. Coarseness, beefiness and anything which interferes with the productiveness and general utility of the breed.

Bantams

Welsummer bantams are to be miniatures of the large fowl and so the standard for large applies.

Weights

Male 1020 g (36 oz)
Female 790 g (28 oz)

Wyandotte

Large fowl

Origin: America
Classification: Heavy
Egg colour: Tinted

The first variety of the Wyandotte family was the silver laced, originated in America, where it was standardized in 1883. The variety was introduced into England at the time, and our breeders immediately perfected the lacings and open ground colouring.

Partridge Cochin and gold spangled Hamburgh males were crossed with the silver females, to produce the gold laced variety. The white Wyandotte came as a sport from the silver laced; the buff followed by crossing buff

Wyandotte

Cochin with the silver laced. In 1896 the partridge variety was introduced from America, the result of blending partridge Cochin and Indian Game blood with that of the gold laced, the variety being perfected for markings in England. It was once called the gold pencilled, and the silver pencilled soon followed from partridge Wyandotte and dark Brahma crossings.

Columbians were the result of crossing the white Wyandotte with the barred Rock, and it was the crossing of the gold laced and the white varieties which produced the buff laced and the blue laced, first seen here in 1897. Blacks, blues and barred have been made in different ways in this country. The latest variety to be introduced is the red, created in Lancashire, from the gold laced variety, with selective matings with white Wyandotte, Barnevelder and Rhode Island Red.

It is clear that while the family of the Wyandotte is large, every variety is a made one from various blendings of breeds.

General characteristics: male

Carriage: Graceful, well balanced, alert and active, but docile.

Type: Body short and deep with well-rounded sides. Back broad and short with full and broad saddle rising with a concave sweep to the tail. Breast full, broad and round with a straight keel bone. Wings of medium size, nicely folded to the side. Tail medium size but full and spread at the base, the main feathers carried rather upright, the sickles of medium length.

Head: Short and broad. Beak stout and well curved. Eyes intelligent and prominent. Comb rose, firmly and evenly set on head, medium in height and width, low, and square at front, gradually tapering towards the back and terminating in a well defined spike (or leader) which should follow the curve of the neck without any upward tendency. The top should be oval and covered with small and rounded points; the side outline being convex to conform to the shape of the skull. Face smooth and fine in texture. Ear-lobes oblong, wattles medium length, fine in texture.

Neck: Of medium length and well arched with full hackle.

Legs and feet: Thighs of medium length, well covered with soft feathers; the fluff fairly close and silky. Shanks medium in length, strong, well rounded, good quality, and free of feather or fluff. Toes, four, straight and well spread.

Plumage: Fairly close and silky, not too abundant or fluffy.

Female

The general characteristics are similar to those of the male, allowing for the natural sexual differences.

Colour

The barred
Male and female plumage: Similar to that of the barred Plymouth Rock.

Wyandotte

Wyandotte large fowl Silver laced female

The black
Male and female plumage: Black with beetle green sheen, undercolour as dark (black) as possible.

The blue
Male and female plumage: One even shade of blue, light to dark, but medium preferred; a clear solid blue, free from mealiness, 'pepper', sandiness, or bronze, and quite clear of lacing; a 'self colour' in fact.

The blue laced
Male plumage: Bay (red-brown or chestnut) with blue markings. Hackles distinctly striped with blue down the centre of each feather and free from black tips or black around the edging. Back and shoulders free from black or smutty blue. Wing bar laced blue, well defined. Breast regularly and distinctly laced from the throat to the back of the thighs and free from double, outer, black or smutty marking. Fluff blue, powdered with gold. Tail solid blue, free from black or white.
Female plumage: Hackle and tail as in the male. Remainder as on the male's breast, the lacing extending to back of thighs into the fluff.

Wyandotte

The buff
Male and female plumage: Clear, sound buff throughout to skin, allowing greater lustre on the hackles and wing bow of the male. With these exceptions the colour should be perfectly uniform, but washiness or a red tinge, mealiness or 'pepper' to be avoided.

The buff laced
Male and female plumage: Rich buff with white markings. Hackles distinctly striped with white down the centre of each feather. Tail, fluff and undercolour white; wing secondaries, white inner web, outer web buff laced. The male's back, shoulders and wing bow rich solid buff. Remainder (in both sexes) clearly and regularly laced with white, the lacing of the female's cushion sometimes continuing into the tail coverts.

The columbian
Male and female plumage: Pearl white with black markings; primaries (wing) black or black edged with white, secondaries black inner web and white outer. The male's neck hackle broadly striped with intense black down the centre of each feather, such stripe entirely surrounded by a clearly-defined white margin and finishing with a decided white point (free from black outer edging or black tips); saddle hackle white, and the tail glossy green-black, coverts either laced or not with white. The female's hackle bright intense black, each feather entirely surrounded by a well defined white margin, and tail feathers black, except the top pair which may or may not be laced with white. Remainder (in both sexes) white, entirely free of ticking, with slate, blue-white or white undercolour.

The gold laced
Male plumage: Head a rich golden bay, the neck hackle with a distinct black stripe down the centre of each feather, free from ticks, black outer edging or black tips. Saddle hackles to match neck. Back a rich golden bay, free from black or maroon. Shoulder tip bay laced with black. Wing bow bay, coverts evenly laced, forming at least two well-defined bars; secondaries black on inner and wide bay stripe on outer web, the edge laced with black; primaries black on inner web and broadly laced bay on outer edge. Breast and underparts, the web bay, with well-defined jet black lacing, free from double or bay outer lacing, the lacing regular from throat to back of thighs and showing a green lustre. Undercolour dark slate. Tail, the true tail feathers, sickle and coverts black showing a green lustre. Thighs and fluff black or dark slate, with clear lacing round hocks and outer side of thighs. Brightness and uniformity of colour to be considered of more value than any particular shade. Gold shaft permissible.
Female plumage: Head and neck hackle as in the male. Breast and back, undercolour dark slate, web bay with regular well-defined jet black lacings, free from double or outer lacing, and showing a green lustre. Wings as the broad portion of the back; secondaries and primaries as in the male. Tail black, showing a green lustre, the coverts black with a bay centre to each feather. Thighs and fluff black or dark slate. Brightness and a quality of ground colour and regularity of lacing throughout are of first importance. Gold shaft permissible.

Wyandotte

Wyandotte large fowl
Columbian male
Columbian female

Wyandotte

Wyandotte large fowl
White male
White female

Wyandotte

Wyandotte large fowl
Partridge male
Black female

Wyandotte

The silver laced
Male and female plumage: Except that the ground is silver white (free from yellow or straw tinge), instead of rich golden bay, the silver laced is similar to the gold laced. In the female, regularity of lacing and quality of colour must count above any particular breadth of lacing. Silver shaft permissible.

The partridge
Male plumage: Head dark orange. Hackles bright orange-yellow, shading to bright lemon yellow, free from washiness, each feather having a clearly-defined glossy black stripe down the middle, not running out at the tip, and free from light shaft. Back and shoulders bright red of a scarlet shade, free from maroon or purple tint. Wing bar solid, glossy black; primaries solid black, free from white; secondaries, rich bay outer web and black inner and end of feather, the rich bay alone showing when the wing is closed. Undercolour black or dark grey, free from white. Breast and fluff metallic black, free from red or grey ticking. Tail (including sickles and tail coverts) metallic black, free from white at roots.

Female plumage: Head and hackle rich golden yellow, the larger feathers finely and clearly pencilled. Breast, back, cushion and wings soft light partridge brown, quite even and free from red or yellow tinge, each feather plentifully and distinctly pencilled with black, the pencilling to follow the form of the feather, and to be even and uniform throughout. Fine, sharply-defined pencilling with three or more distinct lines of black is preferred to coarse, broad marking, especially in females, in which the pencilling is generally better defined than in pullets. Pencilling that runs into the brown, peppery markings, and uneven, broken or barred pencilling, constitute defects. Light shafts to feathers on the breast must be penalized. Fluff brown (same shade as body), as clearly pencilled as possible. Primaries (wing) black, secondaries brown (same shade as body), pencilled with black on outer web, black on inner web, showing pencilling when the wing is closed. Tail black, with or without brown markings, with clearly-pencilled feathers up to the point of the tail.

The silver pencilled
Male and female plumage: Except that the ground is silver white in the male and steel grey in the female, instead of red, brown, etc. (of various shades), the silver pencilled is similar to the partridge.

The red
Male and female plumage: Surface rich, bright, glossy red. Neck hackle of medium shade to match body colour, with a black stripe down the centre of each feather at the lower part. Tail and coverts green-black. Wing primaries, inner half black, outer half red; secondaries, inner half dark slate or black, outer half to match body. Undercolour dark or slate, clearly defined.

The white
Male and female plumage: Pure white, free from yellow or straw tinge.

In both sexes and all colours: Beak bright yellow (except in buff laced, yellow or horn tipped with yellow; columbian and red, yellow or horn; gold laced, partridge, silver laced, and silver pencilled, horn shading into or tipped with yellow). Eyes bright bay (in the red, red or orange). Comb, face, wattles and ear-lobes bright red. Legs and feet bright yellow.

Weights

Cock 3.60–4.10 kg (8–9 lb); cockerel 2.95–3.40 kg (6½–7½ lb)
Hen 3.20 kg (7 lb); pullet 2.50 kg (5½ lb)

Scale of points

The black
Colour (surface 25, undercolour 10)	35
Type	25
Head	10
Size and condition	15
Legs	15
	100

The blue
Type	25
Colour	25
Head	15
Legs	15
Size	10
Condition	10
	100

Serious defects: These include black legs devoid of yellow.

The blue laced
	Male	Female
Colour and markings	64	73
Head	19	10
Size and condition	12	12
Legs	5	5
	100	100

Serious defects: These include white in tail.

The buff
Type (back 10, body 12, wings 10, tail 8)	40
Head 6, comb 8, ear-lobes and wattles 8	22
Size and condition	20
Neck	10
Legs	8
	100

Wyandotte

The buff laced

Colour and markings	70
Head	14
Size and condition	11
Legs	5
	100

Serious defects: These include black in tail, or excess of blue or grey in lacing.

The columbian

Type or shape	25
Comb	10
Eye	5
Body colour	15
Hackle, including scantiness	10
Tail	5
Flights	5
Legs	5
Texture	5
Condition (including activity)	7
Size	8
	100

Serious defects (which should be heavily penalized): These include badly crooked breast bone, coarseness. Inactivity. Excess of feather. Overhanging eyebrows. Crooked toes. Brown undercolour. Green eyes.

The gold or silver laced

Comb and head	10
Ear-lobes and wattles	4
Neck	8
Breast and thighs	12
Back	12
Tail	6
Wings	12
Fluff	5
Legs	5
Size and condition	14
Shape	12
	100

Serious defects: These include white in tail. Any conspicuous spotting or peppering of ground colour.

Wyandotte

The partridge, male

Colour and markings (colour of hackles 8, striping of hackles 8, top colour 8, breast 7, flights 5, tail 4, undercolour 4, fluff 4)	48
Head (comb 7, eyes 5, lobes and wattles 4)	16
Type	22
Size and condition	8
Legs	6
	100

The partridge, female

Colour and markings (ground colour 13, formation – breadth and form of black marks – of pencilling 11, clearness of pencilling 10, fluff 5, hackles 4)	43
Head (comb 6, eyes 5, lobes 4)	15
Type	22
Legs	10
Size and condition	10
	100

Serious defects: These include slipped wings, wall eyes or eyes that do not match.

The red

Indications of egg producing merits	15
Indications of reproductive merits	15
Type and carriage	20
Colour and markings	20
Head (including eyes)	10
Legs	10
Condition	10
	100

Serious defects: These include shanks other than yellow (allowance made for adult birds and heavy laying females). Coarseness, superfine bone. Any points against egg production, reproduction, or stamina values. Absence of any dark undercolour, and any deformity for which a bird may be passed.

The silver pencilled

	Male	*Female*
Colour and markings	48	43
Head	16	15
Size and condition	8	10
Legs	6	10
Type	22	22
	100	100

The division of points in the silver pencilled is precisely the same as in the partridge.

The white

Type	25
Colour	25
Head	15
Legs and feet	10
Size	15
Condition	10
	100

Serious defects (for which a bird may be passed): These include feathers other than white in colour. Coarseness and 'Orpington' type.

Serious defects in all varieties

Any feathers on shanks or toes. Permanent white or yellow in ear-lobe covering more than one-third of its surface. Comb other than rose, or falling over one side, or so large as to obstruct the sight. Shanks other than yellow, except in mature birds, which may shade to light straw. Any deformity.

Bantams

Wyandotte bantams are miniatures of the large fowl and the standards in every respect are the same, with the exception of weights and scale of points. White-laced buffs, violet laced, blue laced, buff and cuckoo are also seen in bantams.

Colour

In both sexes: Beak bright yellow, except in marked and laced varieties, in which it may be horn, shaded with yellow. (*Note:* yellow beaks are unobtainable in black males with dark undercolour, and beak colour in these should be black, shaded with yellow.) Eyes bright bay in all colours. Comb, face, wattles and ear-lobes bright red. Legs and feet bright yellow.

Weights

Male 680–790 g (24–28 oz)
Female 570–680 g (20–24 oz)

Scale of points

The white

Colour	25
Type	25
Head	5
Comb	10
Lobes	5

(*continued*)

(**Scale of points** – *continued*)
Eyes	5
Leg colour	5
Size	10
Condition	10
	100

The black
Colour	20
Undercolour	15
Type	20
Head	5
Comb and lobes	10
Legs colour	10
Size	12
Condition	8
	100

The partridge and silver pencilled, male
Hackle colour	8
Striping of hackles	8
Top colour	8
Breast	7
Flights	5
Tail	4
Undercolour	4
Fluff	4
Comb	7
Eyes	5
Lobes and wattles	4
Type	22
Legs and feet	6
Size and condition	8
	100

The partridge and silver pencilled, female
Ground colour	13
Form of pencilling	11
Clearness of pencilling	10
Fluff	5
Hackle	4
Comb	6
Eyes	5
Lobes	4
Type	22
Legs and feet	10
Size and condition	10
	100

Wyandotte

**Wyandotte bantams
Columbian male
Silver-laced female**

Wyandotte

Wyandotte bantams
Partridge male
White female

Yokohama

The gold or silver laced

Comb and head	10
Lobes and wattles	4
Neck	8
Breast and thighs	12
Back	12
Tail	6
Wings	12
Fluff	5
Legs	5
Size and condition	14
Shape	12
	100

The columbian

Comb	10
Eyes	5
Lobes and wattles	5
Hackle and tail	15
Body colour	15
Legs	5
Type and symmetry	35
Condition	10
	100

Serious defects

Feathers on shanks or toes. Permanent white or yellow in ear-lobes, covering more than one-third of the surface. Comb other than rose or flopping or obstructing the sight. Shanks other than yellow. Any deformity. Slipped wings (which should be penalized strongly). Eyes not matching or other than bright bay. Conspicuous peppering on ground colour of laced varieties. Any form of double lacing in laced varieties.

Yokohama

Origin: Japan
Classification: Light
Egg colour: Tinted

In this country the Yokohama, the long-tailed breed of Japanese origin, is known in two varieties:

1. The Yokohama, with a walnut comb, found in two colours, red saddled and white.

Yokohama

Yokohama large fowl Duckwing male

2. The Yokohama with a single or pea comb, found in black-red and gold duckwing.

In Britain, both these breeds have been known, shown and standardized under the name of Yokohama.

General characteristics: male

Carriage: Stylish and pheasant-like.

Type: Body fairly long and deep, full round breast, long back tapering to tail, long wings carried rather low but close to the sides. Tail as long and flowing as possible, with a great abundance of side hangers, the sickle and coverts narrow and hard, and the whole tail forming a graceful curve and carried somewhat low.

Head: Skull small but inclined to be long and tapering. Beak strong and curved. Eyes bright and full of life. Comb single, pea or walnut (in the red saddled, walnut only allowed), small and even. Face of fine texture. Ear-lobes small, oval and almond shape, fitting closely. Wattles round and small.

Yokohama

Neck: Long and furnished with flowing hackle.

Legs and feet: Of medium length, the shanks fine and free of feathers. Toes, four, well spread.

Female

The general characteristics are similar to those of the male, allowing for the natural sexual differences.

Colour

The black-red, duckwings (silver and gold), blue, red
Male and female plumage: As in the corresponding colours in Game fowl.

In both sexes: Beak horn. Eyes ruby red. Comb, face and wattles bright red. Ear-lobes white or red. Legs and feet yellow, willow or slate blue.

The red saddled
Male and female plumage: White and red. Breast and thighs (red in the male and red-buff in the female), with distinct white spangling at the end of each feather. Male's back and wing bow crimson red, the former vignetted into the saddle. Remainder white.

In both sexes: Beak yellow. Eyes ruby red. Comb, face, wattles and ear-lobes bright red. Legs and feet bright yellow.

The white
Male and female plumage: Snow white, free of any straw tinge.

In both sexes: Beak, legs, and feet white or yellow. Eyes bright red. Comb, face, wattles and ear-lobes red.

Weights

Male   1.80–2.70 kg (4–6 lb)
Female 1.10–1.80 kg (2½–4 lb)

Scale of points

Plumage (quality and length of tail 25, and of neck and saddle hackles 20)	45
Type and condition	25
Head	10
Colour	10
Legs	5
Size	5
	100

Serious defects

Yellow or straw feathers in the white. White in face. Other than four toes. Any deformity.

Turkeys

Beltsville White

It is strange that the first English Book of Standards of 1865 incorporated the turkey, without defining colours or varieties. Yet at the first English Show of 1845 the classification was for 'Whites' and 'Any other colour'. The first turkeys to reach England came from Spain about 1524. A Spanish explorer had discovered turkeys on the coast of Cumana, north of Venezuela in 1499, from which place specimens were shipped to Spain in 1500, and some birds reached France in 1516. It has been said also that turkeys reached England from Spain in 1821. Commercially these standards are now obsolete having been superseded by broad-breasted varieties.

Beltsville White

Origin: America

This is a comparative newcomer to turkey breeds, and was developed in America to meet the consumer demand for a smaller type of bird. It rapidly achieved popularity in the States, and although adopted by British breeders has now given way to the Small White.

General characteristics: male

Type: Body very symmetrical, medium length, and broad. Back broad throughout and flat. Breast medium length, broad, and slightly curved. Wings strong and neatly folded. Tail long and straight, carried in line with the back.

Head: Small length and breadth, and caruncviated. Beak strong, curved, and well set. Eyes prominent, bright, and clear. Throat wattle medium in length and of fine texture, conforming to the size of the bird.

Neck: Of medium length in proportion to rest of the body.

Legs and feet: Thighs of medium length and strong. Fluff short and tight. Shanks fine but strong, flat sided, fairly short. Toes, four, straight and strong.

Plumage: Very tightly feathered.

Blue

Colour

Plumage of both sexes: Pure white throughout, free from creaminess, off-white or other tinge; black beard.

In both sexes: Beak light pinkish horn. Eyes, brown iris, black pupil. Face, wattle, and caruncles bright rich red, but changeable to blue and white in the male. Shanks and feet pink and flesh colour. Toes pale horn.

Weights

Cock 8.15–10.00 kg (18–22 lb); cockerel 5.45 kg–7.70 kg (12–17 lb) min. and max.
Hen 4.55–5.45 kg (10–12 lb); pullet 3.60–4.55 kg (8–10 lb) min. and max.

Scale of points

Defects in	Deduct
Condition	10
Head, neck, and wattle	10
Colour	10
Shape and type	55
Legs and feet	15
	100

Serious defects

Crooked or other deformity of the breast bone. Deep breasts with a pronounced knob on point of breast bone. Wry tail. Feathers other than white. Pronounced debeaking.

Disqualification

Any birds exceeding the weight laid down in the standard.

Blue

Origin: America

This breed has been developed on the lines of the American Slate turkey. The feathers of the slate are ashy blue and may be dotted with black in any part of the plumage, whereas the blue standard calls for an even colour.

General characteristics: male and female

These are the same as for the Buff turkey.

Blue

Blue turkey, male
Norfolk Black turkey, male

British White

Colour

Plumage of both sexes: A light or dark shade of sound and even blue, free from black or brown feathers.

In both sexes: Beak, legs and feet slate blue. Eyes dark to black. Face, jaws, wattle and caruncles bright rich red.

Weights

Cockerel 8.15–11.35 kg (18–25 lb)
Pullet 6.35–8.15 kg (14–18 lb)

Scale of points

Defects in	Deduct up to
Shape (type)	25
Head, neck and wattle	15
Legs and feet	10
Weight	10
Colour	35
Condition	5
	100

Serious defects

Presence of black or brown feathers. Crooked breast, wry tail or any other deformity. Pronounced debeaking.

British White

Origin: Great Britain

White turkeys were bred here in Europe from the early times, and the breed has been known also as the White Holland, and even the Austrian. It retains its name of White Holland in America, being standardized there as such, while there, too, producers have now developed a small Beltsville White. To distinguish our own developed strains of the White variety, it is to be called the British White, as bred to our standard.

General characteristics: male and female

Type: Body long, deep and well rounded. Back curving with good slope to tail. Breast broad, full, long and straight. Wings strong and large. Tail long in proportion to body.

Head: Long, broad and carunculated. Beak stong, curved and well set. Eyes bright, bold and clear. Throat wattle large and pendent.

British White

Bronze turkey, male
British White turkey, male

Buff

Neck: Long, curving backward towards tail.

Legs and feet: Thighs long and stout. Fluff short. Shanks large, strong, well rounded and of medium length. Toes, four, straight and strong and well spread.

Colour

Plumage of both sexes: Pure white with black tassel.

In both sexes: Beak white to pale horn. Eyes, iris dark hazel, pupil blue-black. Face, wattle and caruncles bright rich red, but changeable in the male to blue and white. Shanks and feet pink flesh. Toe-nails white to pale horn.

Weights

Cock 12.70–17.25 kg (28–38 lb); cockerel 8.15–12.70 kg (18–28 lb) min. and max.
Hen 7.25–10.00 kg (16–22 lb); pullet 6.35–8.15 kg (14–18 lb) min. and max.

Scale of points

Defects in	Deduct up to
Shape (type)	35
Head, neck and wattle	10
Legs and feet	5
Weight	25
Colour	15
Condition	10
	100

Serious defects

Feathers other than white. Crooked breast. Wry tail. Any other deformity. Pronounced debeaking.

Buff

Origin: America

In their day our breeders developed excellent strains of the Buff turkey, which were often reckoned to be one of the best layers among the various breeds. It was not aptly named as the colour is actually a deep cinnamon brown matching the colour of a dead beech leaf.

General characteristics: male and female

These are the same as those of the British White, with the exception that the shanks are large, fairly long and strong.

Colour

Plumage of both sexes: Deep cinnamon brown. Flights and secondaries white. Tail deep cinnamon brown, edged with white.

In both sexes: Beak light horn. Eyes, iris dark hazel, pupil blue-black. Face, jaws, wattle and caruncles bright rich red. Shanks and toes pink and flesh. Toe-nails light horn.

Weights

Cock 10.00–12.70 kg (22–28 lb); cockerel 7.25–10.45 kg (16–23 lb)
Hen 5.45–8.15 kg (12–18 lb); pullet 3.60–6.35 kg (8–14 lb)

Scale of points

Defects in	Deduct up to
Shape (type)	20
Head, neck and wattle	15
Legs and feet	10
Weight	25
Colour	25
Condition	5
	100

Serious defects

White in tail except in edging. Crooked breast. Wry tail. Pronounced debeaking. Any other deformity.

Cröllwitzer

Origin: Germany

This is one of the most attractive breeds of turkey seen in this country. It has recently been imported from the Continent.

General characteristics: male and female

Type: Body long, deep and well rounded. Back curving with good slope to tail. Breast broad, full, long and straight. Wings strong and large. Tail long in proportion to body.

Cröllwitzer

Head: Long, broad and carunculated. Beak strong, curved and well set. Eyes bright, bold and clear. Throat wattle large and pendent.

Neck: Long, curving backwards towards the tail.

Legs and feet: Thighs long and stout. Fluff short. Shanks large, strong, well rounded and of medium length. Toes, four, straight and strong and well spread.

Colour

Plumage of both sexes: Ground colour white. Neck to be pure white. Each breast feather to finish with a black band with a 1.2 mm silver edge, these black markings generally better defined on females. On the back, shoulder, sides and tail coverts, markings to be stronger with a 2.4 mm white edge. Large tail feathers to have a diagonal black stripe with a broad end. Shoulder butts black with a white end. Wing bow, bar and coverts white with a black end. Primaries grey-black with a white shaft.

In both sexes: Beak flesh to horn. Leg colour flesh or red.

Weights

Cock 12.70–17.25 kg (28–38 lb); cockerel 8.15–12.70 kg (18–28 lb) min. and max.
Hen 7.25–10.00 kg (16–22 lb); pullet 6.35–8.15 kg (14–18 lb) min. and max.

Scale of points

Defects in	Deduct up to
Shape (type)	20
Head, neck and wattle	15
Legs and feet	10
Weight	25
Colour	25
Condition	5
	100

Serious defects

Any brown in plumage. Any black in neck. White edge on back missing.

Minor defects

Silver edge missing on breast feathers considered only a very minor fault.

Mammoth Bronze

Origin: Great Britain

The Mammoth Bronze turkey was first developed here as the Cambridge, the Norfolk Black being used in the crossings, and it became known as the Cambridge Bronze. Later importations of American Bronze turkeys resulted in crossings with the Cambridge Bronze to obtain greater size, until the name American Mammoth Bronze became generally known.

General characteristics: male and female

Type: Body long, deep and well rounded. Back curving with a good slope to the tail. Breast broad, full, long and straight. Wings strong and large. Tail long in proportion.

Head: Long, broad and carunculated. Beak strong and curved and well set. Eyes bright, bold and clear. Throat wattle large and pendent.

Neck: Long, and curving backwards towards the tail.

Legs and feet: Thighs medium and stout. Fluff short. Shanks large, fairly long and strong. Toes, four, straight, strong and well spread.

Colour

Plumage of both sexes: A good metallic bronze throughout, the female to have a slight ticking on the breast. Flights black with a definite white barring. Tail black and brown, with a good broad black band edged with white.

In both sexes: Beak horn. Eyes, iris a dark hazel, pupil blue-black. Face, jaws, wattle and caruncles bright rich red. Shanks and toes black or horn. Toe-nails horn.

Weights

Cock 13.60 kg (30 lb); cockerel 11.35 kg (25 lb) min.
Hen 8.15–11.80 kg (18–26 lb); pullet 6.35–10.00 kg (14–22 lb) min.

Scale of points

Defects in	Deduct up to
Shape (type)	25
Head, neck and wattle	15
Legs and feet	10
Weight	25
Colour	20
Condition	5
	100

Serious defects

Crooked breast. Wry tail. Pronounced debeaking. Any other deformity.

Norfolk Black

Origin: Great Britain

The original turkey importations into this country were darks and, no doubt, the first variety to be developed here was the Norfolk Black in the county from which it derived its name. Specimens were shipped to America and the breed may well be claimed as the first of the domestic varieties of turkey.

General characteristics: male and female

Type: Body fairly long and deep, particularly broad across the shoulders. Back broad and flat between the shoulders. Breast not too long, well rounded, muscular and fleshy. Wings carried lightly. Tail long in proportion to body.

Head: Fairly long and broad, and carunculated. Beak strong curved and well set. Eyes bright, bold and clear. Throat wattle large and pendent.

Neck: Of medium length curving slightly backward with an alert carriage.

Legs and feet: Legs short to medium length and set well apart. Thighs full and thick. Toes, four, straight, strong and well spread.

Colour

Plumage of both sexes: A dense black.

In both sexes: Beak, legs and feet black. Eyes dark to black. Face, jaws, wattle and caruncles bright rich red; but short black feathers on head and face not a fault.

Weights

Cock 11.35 kg (25 lb); cockerel 8.15–10.00 kg (18–22 lb)
Hen 5.90–6.80 kg (13–15 lb); pullet 5.00–5.90 kg (11–13 lb)

Scale of points

Defects in	Deduct up to
Shape (type)	20
Head, neck and wattle	15
Legs and feet	10
Weight	25
Colour	25
Condition	5
	100

Serious defects

Bronze in undercolour or wings. White flecks on thigh or wing feathers. Crooked breast. Wry tail. Pronounced debeaking. Any other deformity.

Other breeds

Among other breeds of turkey occasionally exhibited in this country may be mentioned the lavender or slate, slate or ash blue dotted (but not laced or spangled) with black; and the Italian grey-black, the feathers edged with a clear grey.

Ducks

It is generally accepted that all breeds of ducks, with the exception of the Muscovy, originated from the wild Mallard. This is quite clear with a breed like the Rouen, and some consider that the Black East Indian and the Cayuga originated from sports of the Mallard. It is possible to understand, too, the original white ducks of this country coming as Mallard sports. They may have been developed for body size and table qualities, by domestication and selection, resulting eventually in the Aylesbury as we know it today. In the make-up of the khaki Campbell the wild Mallard also played its part.

The British Waterfowl Association classifies the following as ornamental ducks; the Carolina, Mandarin and all other British or foreign breeds of wild duck.

Aylesbury

Origin: Great Britain

The Aylesbury derives its name from the town of Aylesbury in Buckinghamshire. At the first poultry show of 1845, a class was provided for 'Aylesbury or other white variety', and another for 'Any other variety'. No doubt there were white ducks in this country for centuries before, and from them was developed by judicious selection for table purposes the white Aylesbury, which is today Britain's table breed *de luxe*. Once standardized as a breed it was developed by selecting for its distinctive characteristics, which separated it from all other white breeds of ducks.

General characteristics: male and female

Carriage: Horizontal, the keel practically parallel with the ground.

Type: Body long, broad and very deep, showing a good keel. Back straight, almost flat. Breast full and prominent. Keel quite straight from breast to stern. Wings strong and carried closely to the sides, fairly high but not touching across the saddle. Tail short, only slightly elevated, and composed of stiff feathers, the drake's having two or three well-curled feathers in the centre.

Head: Strong and powerful, with eyes as near the top of the skull as possible. Bill strong and wedge-shaped. When viewed from the side the outline is almost straight from the top of the skull, the head and bill measuring from 15 to 20 cm (6 to 8 in). Eyes full.

Aylesbury

Aylesbury
Drake
Duck

Neck: Curved and strong.

Legs and feet: Legs very strong and short, the bones thick, set to balance a level carriage. Feet straight and webbed.

Plumage: Bright and glossy, resembling satin.

Colour

Plumage of both sexes: White.

In both sexes: Bill pink-white or flesh. Eyes dark. Legs and webs bright orange.

Weights

Drake 4.55 kg (10 lb)
Duck 4.10 kg (9 lb)

Scale of points

Type	10
Size	20
Head and bill	20
Eyes	8
Keel	10
Colour	10
Neck	5
Legs and feet	5
Condition	12
	100

Serious defects

Plumage other than white. Bill other than white or flesh pink. Heavy behind. Any deformity.

Black East Indian

Origin: America

The Black East Indian is described in the first Book of Standards of 1865, but it has had other names such as Buenos Aires, Labrador, and Black Brazilian. With weights of 0.70–0.80 kg (1½–1¾ lb) for the duck and of 0.90 kg (2 lb) for the drake, it might be regarded as the 'bantam' of the duck breeds. Many consider this a black sport from the Mallard.

General characteristics: male and female

Carriage: Lively, smart, symmetrical and clear of the ground from breast to stern.

Black East Indian

Black East Indian drake

Type: Body short and broad. Breast round and prominent.

Head: Neat and round, with high skill. Bill short and fairly broad, well set in a straight line from the tip of the eye. Eyes full.

Neck: Short.

Legs and feet: Legs of medium length, placed midway in the body. Feet straight and webbed.

Plumage: Bright and glossy.

Colour

Plumage of both sexes: A very lustrous intense beetle green black, free from purple or white feathers, a brown or purple tinge being objectionable but not a disqualification.

In both sexes: Bill slate black, olive patches not objectionable. Eyes dark. Legs and webs as black as possible, but becoming more or less orange with age.

Weights

Drake 0.90 kg (2 lb)
Duck 0.70–0.80 kg (1½–1¾ lb)

Scale of points

Type	20
Size	20
Head, bill and neck	15
Legs and feet	5
Colour	30
Condition	10
	100

Serious defects

White and purple feathers. Slipped wing. Dished bill. Any deformity.

Blue Swedish

Origin: Europe

The Blue Swedish duck has long been admired for its striking appearance. Its rich well-laced blue colour, large size, fine length and carriage, with the added bonus of two white flight feathers, make this bird a challenge for any breeder.

General characteristics: male and female

Carriage: From 20 to 25° above the horizontal. Lively and alert.

Type: Body proportioned to dimensions of the back, round, plump, deep, without keel. Breast full, well meated, deep. Abdomen free of bulkiness, but round, full, and capacious. Back flat, straight, about 50% longer than broad. Tail rather short, compact, carried slightly elevated. Wings neatly folded, maintained well up, but not meeting over the back.

Head: Proportionate in size, oval, neat. Bill medium length, straight culmen, attached from top of the eye. Eyes bold, bright.

Neck: Moderately slender, tapering from shoulders, faintly curved.

Legs and feet: Thighs well fleshed and of sufficient length to display the hocks just below the coverts. Shanks moderately short, strong.

Colour

Plumage of both sexes: Head in the drake, dark blue with greenish reflexions. In the duck, same as body colour. Wings same as body colour,

Blue Swedish

Blue Swedish duck

except that the two outer primaries in each member should appear white. Speculums as inconspicuous as possible. Remainder of plumage a uniform shade of medium blue throughout, except for an unbroken, inverted heart-shaped 'bib' (about 7.5 × 10 cm (3 × 4 in) in extent) upon the lower neck and upper breast.

Bill in the drake blue preferred, but greenish blue acceptable. In the duck, rather dull orange-brown, with a darker saddle. Eyes brown. Legs and webs orange in drake, brownish orange in the duck.

Weights

Adult drake 3.60 kg (8 lb); adult duck 3.20 kg (7 lb)
Young drake 3.20 kg (7 lb); young duck 2.70 kg (6 lb)

Scale of points

Type	20
Weight	20
Colour	30
Head, neck and bill	15
Legs and feet	5
Condition	10
	100

Campbell

Origin: Great Britain

It was the wild Mallard that played its part in the make-up of the khaki Campbell, together with blood of the fawn-and-white Runner and that of the Rouen. Introduced in 1901 by its originator Mrs. Campbell of Uley, Gloucestershire, it was her special desire to keep the breed for prolific egg-laying, so that a very elementary standard was at first publicized. In this way the high egg-producing properties of the breed were maintained. The white Campbell came as a sport from the khaki. The dark Campbell was created by Mr. H. R. S. Humphreys in Devon, to make sex-linkage in ducks possible.

General characteristics: male and female

Carriage: Alert, slightly upright and symmetrical, the head carried high, with shoulders higher than the saddle, and the back showing a gentle slant from shoulder to saddle; the whole carriage not too erect but not as low as to cause waddling. Activity and foraging power to be retained without loss of depth and width of body generally.

Type: Body deep, wide and compact, appearing slightly compressed, retaining depth throughout, especially from shoulders to chest and from middle of back through to thighs; broad and well-rounded front. Back wide, flat and of medium length, gently sloping with shoulders higher than saddle. Abdomen well developed at rear of legs, but not sagging; well-rounded underline of breast and stern. Wings closely carried and rather high. Tail short and small, rising slightly, the drake's with the usual curled feathers.

Head: Refined in jaw and skull. Face full and smooth. Bill proportionate, of medium length, depth and width, well set in a straight line with the top of the skull. Eyes full, bold and bright, showing alertness and expression, high in skull and prominent.

Neck: Of medium length, slender and refined, almost erect.

Legs and feet: Legs of medium length, and well apart to allow of good abdominal development; not too far back. Feet straight and webbed.

Plumage: Tight and silky, giving a sleek appearance.

Quality and refinement: While aiming at good body size emphasis should be placed upon quality or refinement in general, i.e. neat bone, sleek silky plumage, smooth face, fine head points, etc. with absence of coarseness and sluggishness.

Colour

The dark
Drake's plumage: Head and neck beetle green. Shoulders, breast, underparts and flank light brown, each feather finely pencilled with dark

Campbell
Khaki drake
Khaki duck

grey-brown, gradually shading to grey at stern close up to the vent, followed by beetle green feathers with purplish tinge up to the tail coverts. Tail feathers dark grey-brown; coverts beetle green with purplish tinge or reflexion, also curled feathers in centre. Wing bow dark grey-brown laced with light brown; bar broad purplish green band, edged with a thin light grey line on each side; flights and secondaries dark brown.

Bill bluish green with black bean-shaped mark at the tip. Eyes brown. Legs and feet bright orange.

Duck's plumage: Head and neck dark brown. Shoulders, breast and flank light brown, each feather broadly pencilled with dark brown, becoming brown towards stern, with lighter outer lacing, followed by beetle green feathers at the rump. Back and wing bow dark brown, outer laced with lighter brown. Wing bar as in drake, but less lustrous. Tail and wing feathers dark brown.

Bill slaty brown with a black bean-shaped mark at the tip. Eyes brown. Legs and feet as near body colour as possible.

The khaki

Drake's plumage: Head, neck, stern and wing bar green-bronze. Remainder of plumage an even shade of warm khaki shading off to lighter khaki towards the lower parts of the breast.

Bill greenish blue, the darker the better. Legs and webs dark orange.

Duck's plumage: An even shade of warm khaki. Head and neck a slightly darker shade of khaki. Breast lightly laced on close examination. Back laced. Wings khaki, sound top and under.

Bill greenish to slaty black. Legs and webs as near the body colour as possible.

The white

Plumage of both sexes: Pure white throughout.

In both sexes: Bill, legs and webs orange. Eyes grey-blue.

Weights

Drake 2.25–2.50 kg (5–5½ lb)
Duck in laying condition 2.00–2.25 kg (4½–5 lb)

Scale of points

Type (shape and carriage)	25
Size and symmetry	10
Head points	10
Legs and feet	5
Colour	25
Quality and refinement	15
Condition	10
	100

Serious defects

The dark: Yellow bill. White bib or white neck ring. Any deformity. Green eggs. Coarseness.

The khaki: Yellow bill. White bib or white neck ring. Streak from eyes in duck. White or light underpart or top of wings in either sex. White in wing bar. Any deformity. Green eggs. Anything that interferes with production.

The white: Excessive weight or coarseness. Flesh-coloured bill. Any deformity.

Cayuga

Origin: America

Many consider that the Cayuga was bred from the Black East Indian along larger lines. The breed takes its name from Lake Cayuga, New York. It was in 1851 that black ducks made their appearance on the lake, and specimens were sent to this country.

General characteristics: male and female

Carriage: Lively, clear of the ground from breast to stern.

Type: Body long, broad and deep. Breast prominent, keel well forward and forming a straight underline from stem to stern. Tail carried well out and closely folded, the drake's having two or three well-curled feathers in the centre.

Head: Large. Bill long, wide and flat, well set in a straight line from the tip of the eye. Eyes full.

Neck: Long and tapering with a graceful curve.

Legs and feet: Legs large, strong boned, placed midway in the body, giving the bird a carriage similar to that of the Rouen. Feet straight and webbed.

Colour

Plumage of both sexes: A very lustrous green-black, free from purple or white, the whole of the back and upper part of wings, the breast, and underparts of body deep black, the wings naturally more lustrous than the rest of the body plumage; a brown or purple tinge is objectionable, although not a disqualification.

In both sexes: Bill slate black, with dense black saddle in the centre, but not touching the sides or coming within 2.5 cm (1 in) of the end, the bean black. Eyes black. Legs and webs dull orange-brown.

Cayuga duck

Weight

Drake 3.60 kg (8 lb)
Duck 3.20 kg (7 lb)

Scale of points

Type	30
Size	20
Neck	5
Tail	5
Head and bill	10
Condition	10
Legs and feet	5
Colour	15
	100

Serious defects

Red or white feathers. Orange-coloured bill. Dished bill. Any deformity.

White crested duck

Crested

Origin: Great Britain

The Crested is one of the less common breeds of duck, somewhat like the Orpington in size and type. In breeding Crested ducks, not all the ducklings have crests, the plain being distinguishable from the crested at birth. Their utility qualities are comparable with those of the Orpington.

General characteristics: male and female

Carriage: Somewhat erect.

Type: Body long, broad and fairly deep, full round breast, long broad back. Strong wings, carried closely. Short tail, similar to that of the Aylesbury.

Head: Long and straight. Crest globular, large, set evenly on the skull. Bill long and broad. Eyes large and bright.

Neck: Rather long, slightly arched.

Legs and feet: Short and strong. Toes straight and connected by web.

Colour

Any colour is permitted.

Weights

Drake 3.20 kg (7))lb
Duck 2.70 kg (6 lb)

Scale of points

Crest	25
Type	25
Size	15
Head and bill	10
Condition	15
Neck	5
Legs and feet	5
	100

Serious defects

Slipped wings. Any deformity.

Decoy

Origin: Great Britain

These birds have have been standardized since the first edition of this book. Emphasis is placed upon alertness in looks and movement.

General characteristics: male and female

Carriage: Carried well, and nearly level from breast to stern.

Type: Body very small and compact, broad and deep.

Head: Small, neat and round, with high skull. Bill short, maximum length 3.1 cm (1¼ in), and broad, set deep into the skull to give a square-looking appearance. Eyes round, full and alert.

Neck: Short.

Legs and feet: Legs short, set midway in the body. Feet straight and webbed.

Decoy

Decoy
Brown-grey drake
White duck

Colour

The white
Plumage of both sexes: Pure white all over, free from sappy yellow colour.
  In both sexes: Bill bright orange-yellow. Eyes dark.

The brown
Plumage of both sexes: Resembles the same colour as the Mallard.
  In both sexes: Drake's bill an olive green, the duck's as dark as possible. Eyes dark.

The pied
Plumage of both sexes: As with brown. Variations on pied markings allowed, but even markings to be encouraged. Some orange allowed on bill.

The blue-fawn
Plumage of both sexes: A dilute of the brown with a distinct blue shading on back, wings and tail. An apricot-coloured breast to be encouraged. Horn-coloured bill.

The silver
Drake's plumage: Black or green head colour, with roan-coloured body. Horn-coloured bill.
Duck's plumage: Mottled body, being fawn, brown and grey on a white background, with a blue wing flash.

In both sexes and all colours: Webs to be orange, but can be slightly darker in the blue-fawn and silver.

Weights

Drake 570–680 g (20–24 oz)
Duck 450–570 g (16–20 oz)

Scale of points

Type	30
Size	20
Colour	20
Head, neck and bill	15
Legs and feet	5
Condition	10
	100

Note: Decoys, especially the white, should be noticeably alert in looks and movement.

Serious defects

Thin-bodied, boat-shaped bodies. Thin bills and flat skulls.

Indian Runner

Origin: Asia

The era of the high egg-laying breeds of ducks started with the introduction of the Indian Runner into this country from Malaya. A ship's captain brought home fawns, fawn-and-whites, and whites, distributing them among his friends in Dumfriesshire and Cumberland. They proved prolific layers, and there was a class of fawn Runners at the Dumfries show in 1876 but the fawn-and-whites were not exhibited until 1896. The Indian Runner Duck Club's Standard of 1907 described only the fawn-and-white, that of 1913 recognized also the fawn, while the 1926 Standard described the black and the chocolate varieties.

General characteristics: male and female

Carriage: Upright and active. The angle of inclination of the body to the horizontal varies from 50 to 80° when the bird is on the move and not alarmed; but when standing at attention, or excited, or specially trained for the show-pen, it may assume an almost perpendicular pose.

Type: Body slim, elongated and rounded, but slightly flattened across the shoulders. At the lower extremity the front line sweeps gradually round to the tail, which is neat and compact and almost in a line with the body or horizontally, but in some excellent birds slightly elevated or tilted upwards – the position of the tail varying with the attitude of the duck; however habitually-upturned sterns and tails (as in the Pekin duck) are considered objectionable. Stern short compared with other breeds, the prominence of the abdomen and stern varying in ducks according to the season and the age of the bird, being fuller when in lay; but a large pendulous abdomen and long stern, or a 'cut-away' abdomen and stern in young ducks should be avoided. Wings small in proportion to the size of the bird, tightly packed to the body and well tucked up, the tips of the long flights of the opposite wings crossing each other over the rump, more particularly when the bird is standing at attention. At the upper extremity the body gradually and imperceptibly contracts to form a funnel-shaped process, which again without obvious junction merges into the neck proper, the lower or thickest portion of this funnel-shaped process or 'neck expansion' being reckoned as part of the body.

Note: Total length of drake 65–80 cm (26–32 in) and duck 60–70 cm (24–28 in). Length of neck proper, from top of skull to where it joins the thick part of the 'funnel' about one-third the total length of the bird, not less. Measurements should be taken with the bird fully extended in a straight line, the bill and head in a line with the neck and body, and the legs and feet in the same straight line, the measurements being from the tip of the bill to the tip of the middle toe.

Head: Lean and racy looking, and with the bill wedge-shaped. Skull flat on top, and the eye socket set so high that its upper margin seems almost to

Indian Runner

Indian runner
White drake

Indian runner
Fawn duck

Indian Runner

project above the line of the skull. Eyes full, bright, very alert and intelligent. Bill strong and deep at the base where it fits imperceptibly into the skull. There should be no indication of a joint or 'stop'. The upper mandible should be very strong and nicely ridged from side to side, and the line of the lower mandible straight also, with no depression or hollow in the upper line from its tip to its base. The outline should run with a clean sweep from the tip of the bill to the back of the skull. The length and depth of the bill varies, but should never be out of balance or harmony with the rest of the head and the lines of the bird as a whole.

Neck: Long and slender, and when the bird is on the move or standing at attention, almost in a line with the body, the head being high and slightly forward. The thinnest part in fawn drakes is approximately where the dark bronze of the head and upper neck joins the lower or fawn of the neck proper. The muscular part should be well marked, rounded and stand out from the windpipe and gullet, the extreme hardness of feather helping to accentuate this. The neck should be neatly fitted to the head.

Legs and feet: Legs set far back to allow of upright carriage. Thighs strong and muscular, longer than in most breeds. Shanks short and feet supple and webbed. There should be sufficient width between the legs to allow of free egg production, but not as much as to cause the duck, on actual test, to roll or waddle when in motion.

Plumage: Tight and hard.

Colour

The black
Plumage of both sexes: Solid black with metallic lustre like the Black East Indian. There should be no grey under the chin or wings, no grey wing ribbons, and no 'chain armour' on the breast.

In both sexes: Bill black. Legs and webs black or very dark tan.

The chocolate
Plumage of both sexes: A rich chocolate throughout, the drake, on assuming male plumage, darker than the duck, but the ground work is the same.

In both sexes: Bill, legs and webs black.

The fawn
Drake's plumage: Head and upper part of the neck dark bronze with metallic sheen, which may show a faint green tinge meeting the colour of the lower part of the neck with a clean cut or the lower colour merging into it imperceptibly. Lower neck and 'neck expansion' rich brown-red continued on to breast, over the top of the shoulders and upwards to where it joins the head and upper neck colour, merging gradually on the back and breast into the body colour. Lower chest, flanks and abdomen French grey, made up of very minute and dense peppering of dark brown, or almost black, dots on a nearly white ground, giving a general grey effect without any show of white, the grey extending beyond the vent until it meets the dark or almost black feathers of the cushion under the tail. Scapulars (the long pointed feathers on each side of the back covering the

Indian Runner

roots of the wings) red-brown, peppered. Back and rump deep brown, almost black. Tail (fan feathers and curl) dark brown, almost black. Wing bow fawn, not pencilled; bar fawn corresponding with the coverts in the lower part, the upper part darker brown corresponding with the secondaries, which are black-brown with slight metallic lustre; primaries brown, fairly dark. When the drake is in 'eclipse' or in duck plumage he more closely approaches the duck in colour. All the dominant colours fade, but his head and neck are darker than the duck's; the body becomes a dirty fawn or ash, with perhaps some rustiness on the breast.

Bill pure black to olive green, mottled with black, and black bean. Legs and webs black, or dark tan, mottled with black.

Duck's plumage: The general plumage colour is an almost uniform warm ginger fawn, with no marked variation of shade but having a slightly mottled or speckled appearance. When closely examined the head, neck, lower part of chest and abdomen may appear a shade lighter than the rest of the body. Each feather of the head and neck has a fine line of dark red-brown, giving a ticked appearance. Lower part of neck and neck expansion a shade warmer, each feather pencilled with a warm red-brown. Scapulars rich ginger fawn, a shade darker than the shoulder and back, with well-marked red-brown pencilling. Wing bow a shade lighter than the scapulars but darkening towards the bar, the feathers pencilled as before; secondaries warm red-brown; primaries a shade lighter. Back and rump darker, the pencilling being richer and more marked, but the ground colour becoming lighter and warmer towards the tail. Tail lighter than upper parts of body, each feather pencilled. Belly lighter than upper parts of body, about the same shade of fawn as the head and neck, becoming a trifle darker on the tail-cushion, all feathers pencilled. In young ducks that have just completed the first adult moult there is often a rosy tinge on the lower neck expansion, upper part of breast and shoulders, but this soon fades away. Some ducks have a cream or light-coloured narrow band in the wing bar owing to the upper part of each feather being of a lighter or almost cream shade edged or laced again with the normal dark shade.

Bill black. Eye iris golden brown. Legs and webs black or dark tan.

The fawn-and-white

Plumage of both sexes: The cap and cheek markings in the duck nearly the same shade of fawn as the body colour, but dull bronze-green in the drake; the cap separated from the cheek markings by a projection from the white of the neck extending up to, and in most cases terminating in, a narrow line more or less encircling the eye. The cap should come round the back of the skull with a clean sweep – there should be no 'tails' to it. The cheek markings should not extend on to the neck. Bill divided from head markings by a narrow prolongation of the neck-white from 0.31 to 0.62 cm (⅛ to ¼ in) wide, extending or projecting from the white underneath the chin. Neck pure white to about where the neck expansion begins, and meeting the body with a clean cut. Body uniform soft warm or ginger fawn to the skin, the undercolour not of a different shade. The rump and tail of the drake, including the under surface of his tail, a similar hue to his head. When closely examined the coloured body feathers of the drake show a soft warm ground, slightly peppered with a rather warm shade, as only the

Indian Runner

outer edge of the feather is visible. The colour seems solid and more ruddy than that of the duck. The duck should have the same shade of fawn as the fawn duck. The fawn and the white should meet on the breast with an even cut about half-way between the point of the breast bone and the legs. The base of the neck, upper part of wings, back and tail should be as nearly as possible the same colour as the fawn of the breast, and from the fawn of the back an irregular branch on either side extending downwards on the thighs to, or nearly to, the hough. The white of the breast extends downwards between the legs to beyond the vent and may overlap the thighs in part; if the bird is coloured between the ribs and thigh it is termed 'foul flanked'. Wing primaries, secondaries, and lower part of bow pure white, which gives the appearance of a 'heart' laid flat on the bird's back.

In both sexes: Bill light orange-yellow in young birds; entirely, or almost entirely, dull cucumber in adult duck and green-yellow in the adult drake. Legs and webs orange-red.

The white
Plumage of both sexes: A pure white throughout.

Bill, legs and webs orange-yellow. Eye iris blue.

The mallard coloured
Drake's plumage: Head and neck emerald green. The lower third is separated from the brown-red breast by a pure white ring round the neck which is not closed at the back (of the neck). The upper back is grey-brown, while the lower back, rump (parson's nose) and tail show a black-green colouring. Wings are likewise of a brown-grey colour, both adorned by a steel blue surface, which are enclosed by black, then narrow white stripes. The underside of the wings appears as grey-white. Underbody and belly region should show a spotless, fine pearl grey with proportionate (even) black waves.

Legs orange-yellow.

Duck's plumage: Basic colour an even brown all over the body. Each feather has a black-brown sharp design. The aim is to achieve the colour of the wild duck. On both sides of the head there are 'stripes' (bridles, reins) from beak to nape of neck. Wing spans an even brown surface, as with the drake.

Legs somewhat darker than with the drake.

The Cumberland blue
Plumage of both sexes: Rich blue with dark shading on each feather, richest on back. The drake's head is a darker blue and the head of the duck is similar in shade to the body plumage which is not as richly shaded as that of the drake.

In both sexes: Bill bluish green in the drake and bluish grey in the duck. Eyes dark brown. Legs and webs smoky orange.

The trout coloured
Drake's plumage: Head and neck green with a white ring round neck. Back and wing silver grey with a blue surface. Belly a light colour, likely (potentially) to be ivory coloured with brown spots.

Duck's plumage: Neck light brown. Back silver grey, and wing light grey with blue surfaces. Breast and belly ivory coloured, and the trunk wholly brown spotted.

In both sexes: Bill orange-yellow with green spots. Feet orange-yellow.

Weights

Drake 1.60–2.25 kg (3½–5 lb)
Duck 1.35–2.00 kg (3–4½ lb)

Birds bred and shown in the same year as hatched may be accepted for competition at 0.25 kg (½ lb) less.

Scale of points

Body – shape and general appearance of, including lower part of neck, legs and feet	35
Carriage and action	30
Head, eyes, bill and neck, exclusive of lower neck expansion	20
Colour and condition	15
	100

Serious defects

Above and below standard weights and measurements. Body squat and short, oval or flattened. Domed skull with central position of eyes. Plump head, arched (domed) brow (face). Bill dished, weak, 'Roman', under-curved or flat. Neck thick and short, swan or curved. Neck expansion too far back on body causing a chesty appearance in front with a hollow behind. Bulky trunk. Legs set too far forward causing poor carriage. Waddling or rolling gait. Natural carriage in any duck below 40°. Long stern. Wry tail. Flattened back. Slipped wing or any deformity. Drawn-in tail, bad colour. In fawns, white anywhere; eyebrows or eye-stripes; light or cream wings (bows, coverts and flights); in the duck, blue or green wing bars; orange or yellow bill, feet or legs.

Note: In the standard issued by the Indian Runner Duck Club of Great Britain appear the following remarks: 'The Indian Runner, compared with the larger domesticated varieties, is a small, hard-feathered duck of very upright carriage, active habits, and moves with a straight-out walk or run, quite distinctive and different from the roll or waddle of the other domesticated ducks. It is a great traveller and forager, being much less dependent on the proximity of deep or swimming water than other ducks, and it is a prolific layer. Its body appears elongated and somewhat cylindrical, the legs being set on very far back. . . . It has long been an axiom of the Indian Runner Duck Club that a duck which cannot maintain a natural carriage of at least 40° to the horizontal will not be considered a pure Runner, however good its other points may be. . . .'

Magpie

Magpie drake

Magpie

Origin: Wales

The Magpie is a very striking duck with its bold black and white or blue and white plumage. It is of course a medium-sized breed which yields both meat and eggs, and is a good bird for the shows.

General characteristics: male and female

General shape and carriage: Fairly broad across and deep; great length of body, giving a somewhat racy appearance, indicative of strength combined with great activity.

Type: Back of great length, level and fairly broad across. Breast full and nicely rounded. Wings powerful, carried close to body. Tail medium length, gently rising from back, and increasing apparent length of bird; the drake having the usual curled feathers. Abdomen well developed.

Head: Long and straight. Eyes large and prominent, giving keen and alert appearance. Bill long, broad and slightly dished.

Neck: Long, strong and nicely curved.

Legs and feet: Medium, set wide apart. Feet straight and webbed.

Colour

Plumage of both sexes: Head and neck white, surmounted by a black cap covering the whole of the crown of the head to the top of the eyes. Breast white. Back solid black from the points of the juncture of wing bows and body in a straight line across the back of the tail and extending over the wings, giving the effect, when looking at the duck from behind, of a heart-shaped black mantle. The point of juncture of black and white should be sharp and clearly defined. Wing primaries white, secondaries white. The black upper part should show a clearly-defined curve. Tail black. Thighs and rump white.

In both sexes: Bill pale yellow to deep orange. Eyes dark grey or dark brown. Legs and webs orange. Black on legs and webs a slight defect.

Weights

Drake 2.50–3.20 kg (5½–7 lb)
Duck 2.00–2.70 kg (4½–6 lb)

Scale of points

Weight	8
Condition	10
Head (bill, eyes)	10
Neck	2
Back	4
Tail	2
Wings	2
Breast	6
Body	20
Legs and feet	4
Colour	32
	100

Serious defects

Wry tail. Crooked back. Slipped or twisted wings. Feet not webbed. Excessive weight. Excessive coarseness.

Muscovy

Origin: America

The Muscovy has also been known as the Musk duck, and the Brazilian. It is a distinct species, and wild ancestors of it were found in South America, which must have credit for the original source. Under domestication, our breeders have greatly increased the size of the ducks, and also perfected

Muscovy

their colour and markings. In more recent years the breed has become very popular, and specimens are seen at most shows where waterfowl are scheduled.

General characteristics: male and female

Carriage: Low and jaunty.

Type: Body broad, deep, very long and powerful. Breast full, well rounded and carried low; keel long and well fleshed, just clear of ground, slightly rounded from stem to stern. Wings very strong, long and carried high. Tail long and carried low to give the body a longer appearance to the eye and a slightly curved outline to the top of the body.

Head: Large, and adorned with a small crest of feathers (more pronounced in the drake than duck) which are raised erect in excitement or alarm. Caruncles on face and over base of the bill. Bill wide, strong, of medium length and slightly curved. Eyes large, with wild or fierce expression.

Neck: Of medium length, strong and almost erect.

Legs and feet: Legs strong, wide apart and fairly short. Thighs short, strong and well fleshed. Feet straight, webbed, with pronounced toe-nails.

Plumage: Close.

Handling: Hard, well fleshed, muscular.

Colour

The white-winged black
Plumage of both sexes: A dense black throughout, except white wing bows. The black to carry a metallic green sheen, or lustre, with bronze on the breast and parts of the neck.

The white-winged blue
Plumage of both sexes: Blue except for white wing bows.

The black
Plumage of both sexes: A dense beetle green black throughout, with bronze on the breast and parts of the neck.

The white
Plumage of both sexes: Pure white throughout.

The blue
Plumage of both sexes: Light or dark shade of blue.

The black and white
Plumage of both sexes: Black and white, with defined regularity of markings.

The blue and white
Plumage of both sexes: Blue and white, with defined regularity of markings.

Muscovy duck

Variations in colour: There are colour variations according to the countries of importation. In the black and white, also the blue and white, it is customary in some countries for the black or blue to predominate in winning specimens at the shows.

In both sexes and all colours: Bill yellow and black, red, flesh or lighter shade at point. Face and caruncles red or black. Eyes from yellow and brown to blue. Legs and webs white to black.

Weights

Drake 4.55–6.35 kg (10–14 lb)
Duck 2.25–3.20 kg (5–7 lb)

It is a characteristic of the breed for the drake to be about twice the size of the duck.

Scale of points

Shape and carriage	40
Head (including crest and carunculations)	20
Size	20
Condition	10
Colour	10
	100

Orpington

Origin: Great Britain

It was from the blending of the Indian Runner, Rouen and Aylesbury that Mr. W. Cook in Kent made the buff Orpington, intending it to be a dual-purpose breed. Its introduction followed that of the khaki Campbell, and it has been said that the originator was trying to make a strain of khaki duck. At one time very popular for its high laying qualities combined with table properties and also its beauty of plumage and colouring, it has lost ground of late years.

General characteristics: male and female

Carriage: Slightly elevated at the shoulders, i.e. not quite as horizontal as the Aylesbury, but avoiding any tendency to the upright carriage of the Pekin or Indian Runner.

Type: Body long and broad, deep, without any sign of keel. Back perfectly straight. Breast full and round. Wings strong and carried closely to the sides. Tail small, compact and rising slightly from the line of the back, the drake's having two or three curled feathers in the centre. When in lay the duck's abdomen should be nearly touching the ground.

Head: Fine and oval in shape, with narrow skull. Bill of moderate length, the upper mandible straight from bean to base in line with the highest point of the skull. Eyes large and bold, set high in the head. Deep set and scowling eyes are most objectionable.

Neck: Of moderate length, slender and upright.

Legs and feet: Legs of moderate length, strong and set well apart. Feet straight and webbed.

Plumage: Tight and glossy.

Colour

Drake's plumage: A rich even shade of deep red-buff throughout, free from lacing, barring and pencilling except that head and neck are seal brown with bright gloss, but complete absence of beetle green, the seal brown to terminate in a sharply-defined line all the way round the neck. The rump red-brown, as free as possible from 'blue'.

Duck's plumage: Similar to that of the drake's body, and free from blue, brown or white feathers.

In both sexes: Bill orange with dark bean. Eyes brown iris and blue pencil. Legs and webs bright orange-red.

Weights

Drake 2.25–3.40 kg (5–7½ lb) min. and max.
Duck 2.25–3.20 kg (5–7 lb) min. and max.

Orpington Buff duck

Scale of points

Type and size	40
Head and eyes	10
Colour	40
Condition	10
	100

Serious defects

Colour other than stated. Twisted wings, wry tail, humped back or any other physical deformity. In the drake, grey, silver, or blue head, white feathers in neck, brown secondaries, beetle green on any part, very green bill. In the duck, very heavy lacing, strong light line over the eyes, white feathers on neck or breast, brown feathers, green bill.

Pekin

Origin: Asia

Bred in China, the Pekin reached this country around 1874, and about the same time stock also went to America. English breeders called for a plumage of 'buff canary, sound and uniform, or deep cream, the former preferred'. With judges showing preference for the 'buff canary' plumage, the breed died out. In America, however, the Pekin became the producer

Pekin

Pekin Duck

of high-class table ducklings, with the standard plumage of 'creamy white'. Today there are a number of breeders of Pekins of this type, and a change in the new standard in regard to plumage colouring may help to popularize the breed.

General characteristics: male and female

Carriage: Almost upright, elevated in front and sloping downwards to rear.

Type: Body broad and of medium length and without any indication of keel except a little between the legs. Back broad. Breast broad and full followed in underline by the keel (which shows very slightly between the legs) to a broad, deep paunch and stern carried just clear of the ground. Wings short, carried closely to the sides. Tail well spread and carried high, the drake's having two or three curled feathers on top. A good description of the general shape of the Pekin is that it resembles a small wide boat standing almost on its stern, and the bow leaning slightly forward.

Head: Large and broad and round, with high skull, rising rather abruptly from the base of the bill, and heavy cheeks. Bill short, broad and thick, slightly convex but not dished. Eyes partly shaded by heavy eyebrows and bulky cheeks.

Neck: Long and thick, carried well forward in a graceful arch or curve and with slightly-gulleted throat.

Legs and feet: Legs strong and stout, set well back and causing erect carriage. Feet straight and webbed.

Plumage: Very abundant, thighs and fluff well furnished with long, soft downy feathers.

Colour

Plumage of both sexes: Cream or creamy white.
 In both sexes: Bill bright orange and free from black marks or spots. Eyes dark lead blue. Legs and webs bright orange.

Weights

Drake 4.10 kg (9 lb)
Duck 3.60 kg (8 lb)

Scale of points

Type	25
Size	20
Head (bill 10, eyes 5)	15
Neck	5
Colour	15
Tail	5
Legs and feet	5
Condition	10
	100

Serious defects

Black marks or spots on bill. Any deformity.

Rouen

Origin: France

Confirmation of the Mallard as the progenitor of duck breeds is found in the Rouen, which so closely resembles it in plumage markings, while the Rouen drake moults into duck plumage in the summer like the wild Mallard drake. The Rouen undoubtedly came from Rouen in France, and was known also as the Rhône duck. When first brought to England from France the breed was developed for table properties and was used in table crossings. Later it was bred, as today, for beauty of plumage and markings.

General characteristics: male and female

Carriage: Carried horizontal, the keel parallel with ground and just clear of it.

Type: Body long, broad and square. Keel deep, just clear of the ground from stem to stern. Breast broad and deep. Wings large and well tucked to the sides. Tail very slightly elevated, the drake's having two or three curled feathers in the centre.

Rouen Drake
Duck

Head: Massive. Bill long, wide and flat, set on in a straight line from the tip of the eye. Eyes bold.

Neck: Long, tapering and erect, slightly curved but not arched.

Legs and feet: Legs of medium length. Shanks stout and well set to balance the body in a straight line. Feet straight and webbed.

Colour

Drake's plumage: Head and neck rich iridescent green to within about 2.5 cm (1 in) of the shoulders where the ring appears. Ring perfectly white and cleanly cut, dividing the neck and breast colours, but not quite encircling the neck, leaving a small space at the back. Breast rich claret, coming well under; cleanly cut, not running into the body colour, quite free from white pencilling or chain armour; chain armour or flank pencilling rich blue French grey, well pencilled across with glossy black, perfectly free from white, rust or iron. Stern same as flank, very boldly pencilled close up to the vent, finishing in an indistinct curved line (perfectly free from white) followed by rich black feathers up to the tail coverts. Tail coverts black or slate black with brown tinge, with two or three green-black curled feathers in the centre. Back and rump rich green-black from between the shoulders to the rump. Wings: large coverts pale clear grey; small coverts French grey very finely pencilled; pinion coverts dark grey or slate black; bars (two, composed of one line of white in the centre of the small coverts) grey tipped with black, also forming a line at the base of flight coverts, the latter feathers slate black on the upper side of the quill and rich iridescent blue on the lower side, each of these feathers tipped with white at the end of the lower side, forming two distinct white bars (the pinion bar being edged with black) with a bold blue ribbon mark between the two, each colour being clear and distinct and making a striking contrast; flights slate black with brown tinge free from white. Markings throughout the whole plumage should be cleanly cut and well defined in every detail, the colours distinct and not shading into each other.

Bill bright green-yellow, with black bean at the tip. Eyes dark hazel. Legs and webs bright brick red.

Duck's plumage: Head rich (golden, almond or chestnut) brown with a wide brown-black line from the base of the bill to the neck, and very bold, black lines across the head, above and below the eyes, filled in with smaller lines. Neck the same colour as the head, with a wide brown line at the back from the shoulders, shading to black at the head. Wing bars, two distinct white bars with a bold blue ribbon mark between, as in the drake; flights slate black with brown tinge, no white. Remainder of plumage rich (golden almond or chestnut) brown of level shade, every feather distinctly pencilled from throat and breast to flank and stern, the markings to be rich black or very dark brown, the black pencilling on the rump having a green lustre.

Bill bright orange, with black bean at tip, and black saddle extending almost to each side and about two-thirds down towards the tip. Eyes dark hazel. Legs and webs dull orange-brown.

Rouen Clair

Weights

Drake 4.55 kg (10 lb)
Duck 4.10 kg (9 lb)

Scale of points

Drake
Type	10
Size	10
Head	5
Legs and feet	5
Colour (breast 10, bill 5, neck 5, chain armour 10, back and rump 5, wings 5, stern 5, tail 5)	50
Markings	10
Condition	10
	100

Duck
Type	10
Size	10
Head	5
Legs and feet	5
Colour (ground 15, bill 10, head 5, neck 5, wings 5)	40
Pencilling	20
Condition	10
	100

Serious defects

Leaden bill. No wing bars. White flights. Stern broken down. Wings down or twisted. Any deformity. In the drake, black saddle, black bill or minus ring (on neck); in the duck, white or approaching white ring (on neck).

Rouen Clair

Origin: France

It must remind one of the Common Mallard – a body long and developed, the width corresponding with the length.

General characteristics: male and female

General shape and carriage: Very long, graceful with good width. The important feature of the Rouen Clair is its length (ideally 90 cm (35 in) from point of beak to end of tail with neck extended), which gives it elegance in spite of its heavy body.

Rouen Clair

Type: Rather more upright and smooth breasted than the Rouen.

Head: As for Rouen.

Neck: As for Rouen.

Legs and feet: As for Rouen.

Colour

Drake's plumage: Head and neck green with clear white collar, covering about four-fifths of the neck – no grey in the green. Breast (bib) red chestnut, with light white borders at the end of each feather. Belly light grey, fading to white, without reaching the underneath of the tail, which is black. Back, the above pearly grey, darker than the flanks. Wings violet blue (indigo) with brilliant reflecting powers, the ends of which above and below limited by a white streak. Sides (flanks) pearly grey – no mixture of chestnut feathers. Croup, the above brilliant black. Tail feathers whitish grey, garnished with some curled-up feathers, brilliant black (outwardly sign of the drake).

Bill yellow, with a slight greenish tint and no black lines along its centre. Eye yellow. Legs and webs yellow-orange.

Duck's plumage: Wings to be the same as the drake with the brilliant reflexions. Eyebrows, the above of the eye presenting a slight curve nearly white which forms the eyebrow. On the cheek another white line going from the eye to the start of the bill. Feathers, the ground of the plumage to be of the tone '*Isabelle clair*', and dull for each feather of the back, sides and body. Each back feather must have this '*Isabelle*' background with a brown mark in the shape of a horseshoe, or more exactly, a chevron, lightly rounded at the intersection of the two lines, which opens in the shape of a 'V'. The top of the bill and the front of the neck are of a pale colour (creamy shade extending not too low on the breast).

Bill yellow ochre (light transparent green). Eye, pupil blue, iris very dark brown. Legs and webs yellow-orange.

Weights

Drake 3.40–4.10 kg (7½–9 lb)
Duck 3.00–3.40 kg (6⅞–7½ lb)

Scale of points

Size	8
Condition	10
Head, eyes and bill	8
Neck	6
Back	4
Breast	6
Belly	2
Tail	2

(*continued*)

Rouen Clair

Rouen Clair
Drake
Duck

(**Scale of points** – *continued*)
Wings	2
Legs	2
Feet	2
Breed character	10
Colour	38
	100

Saxony

Origin: Germany

The Saxony duck made its first public appearance in the Saxony County Show of 1934. It is a dual-purpose domestic duck, attractive in appearance, producing full-breasted meaty birds and not less than 150 eggs per year.

General characteristics: male and female

Carriage and appearance: Strong, with a long broad body and no trace of keel.

Type: Body long and stocky. Back broad and long, sloping slightly to the rear. Breast broad and deep without a keel. Wings not too long, lying into the sides. Tail long, carried full and closed.

Head: Long and flat. Bill average length and broad.

Neck: Average length, not thin.

Legs and feet: Average length, set almost in the middle of the body.

Plumage: Lies close into the body. Underfeathers soft down, light coloured.

Colour

Drake's plumage: Head and neck blue as far as the white neck ring. Lower part of neck, shoulders and breast rusty red with slight silver lacing on breast. Back and rump blue-grey. Tail feathers and wings oatmeal, wing bars blue.

Duck's plumage: Head, neck and breast buff with a white eye line. There is a suggestion of a light neck ring, broken at the back. Back and breast paler buff. Wing bars and tail a light blue shade. Wings cream.

In both sexes: Bill yellow – plain or with pale green shading in the drake and browny in duck. Eyes dark brown. Legs and webs dark yellow.

Saxony

Saxony drake

Weights

Drake 3.60 kg (8 lb)
Duck 3.20 kg (7 lb)

Scale of points

Type	20
Size	20
Head, bill and neck	15
Eyes	5
Colour	25
Legs and feet	5
Condition	10
	100

Serious defects

Broken neck ring in the drake. Pale eyes. Upright walk. Slipped wing. Coloured nostrils. Dark underfeathering. Brown head on the drake. White bib.

Silver Appleyard

Origin: Great Britain

The Silver Appleyard is a good all-round utility duck produced by Mr. Reginald Appleyard from selective cross breeding. It is a good layer, an excellent table bird and very ornamental.

General characteristics: male and female

Carriage: Lively, slightly erect, the back sloping gently from shoulder to tail.

Type: Body compact, broad and well rounded. Tail broad, the drake having the usual curled tail feathers.

Head: Fine and alert.

Neck: Upright and of medium length.

Legs and feet: Grey-fawn.

Colour

Drake's plumage: Head and neck are black-green with silver white throat flecked with fawn. Silver white ring completely circling the base of the neck. Base of neck and shoulders below the ring light claret. Silver light wing coverts matching the breast and underbody, followed by the usual band of iridescent blue. Back and rump black-green with white tips to tail feathers.

Duck's plumage: Head, neck and underbody silver white with crown and back of neck flecked with fawn. Deep fawn line through eyes. Shoulders and back strongly flecked with fawn, with the usual iridescent blue in the wings. Tail fawn.

In both sexes: Bill yellow. Eyes dark hazel. Legs light orange.

Weights

Drake 3.60–4.10 kg (8–9 lb)
Duck 3.20–3.60 kg (7–8 lb)

Scale of points

Type	30
Colour	25
Size	20
Carriage	15
Condition	10
	100

Silver Appleyard drake (image caption in picture)

Silver Appleyard Bantams

Origin: Great Britain

The bantam is a variation on the large Appleyard, with similar colouring.

General characteristics: male and female

Carriage: Carried well, nearly level from breast to stern.

Type: Body small, the shape bearing a strong resemblance to the Mallard. (A more slender shape than the Decoy.)

Head: Small, neat and slender. Bill medium length and width. Eyes round, full and alert.

Neck: Medium length.

Legs and feet: Legs short, set midway in the body. Feet straight and webbed.

Weights

Drake 680–790 kg (24–28 oz)
Duck 570–680 kg (20–24 oz)

Silver Appleyard bantam Duck

Scale of points

Type	30
Colour	25
Size	20
Carriage	15
Condition	10
	100

Welsh Harlequin

Origin: Great Britain

The Welsh Harlequin is a utility duck which happens to be attractive in appearance. It is a prolific layer, as would indicate its origin from two sports off a khaki Campbell flock in 1949.

General characteristics: male and female

Carriage: Sprightly, slightly erect, head held high, the back sloping gently from shoulder to saddle. Keel well clear of the ground, for active foraging.

Type: Body compact, with width well maintained from stem to stern. Tail short and small. Drakes have the usual curled tail feathers.

Welsh Harlequin

Welsh Harlequin
Drake
Duck

Head: Fine-drawn and held high.

Neck: Almost vertical, medium in length.

Colour

Drake's plumage: Head and neck bottle green to within about 2.5 cm (1 in) of the shoulders, where a ring 1.25 cm (½ in) wide, completely encircles the neck. In all other respects, except in size, the drake is coloured just like the Mallard.

Bill gun metal coloured. Legs orange.

Duck's plumage: Head and neck fawn. The rest of the body basically cream, with tortoise-shell effect in red-brown and blue on back and wings. Bars on wings are electric blue.

Bill pale yellow or khaki. Legs dark brown.

Weights

Drake 2.50–2.70 kg (5½–6 lb)
Duck 2.25–2.50 kg (5–5½ lb)

Scale of points

Type	20
Size	15
Head, bill and neck	15
Carriage	10
Colour	15
Keel	5
Legs and feet	5
Condition	15
	100

Other breeds

Below are mentioned two breeds of ducks that are more or less moribund if not extinct, although such a statement must be made with reserve as there are probably specimens of some of these breeds still in existence. The British Waterfowl Association looks after the other breeds of duck not standardized in this edition.

Bali: Origin: East India. Type: Head resembling that of a Runner, as also shape of body with a crest mainly globular. Wings closely packed. Carriage erect. Weights: Drake 2.25 kg (5 lb) min; duck 1.80 kg (4 lb). Colour: Plumage white. Bill orange-yellow. Eyes iris blue. Legs and webs orange-yellow.

Penguin: Origin: Great Britain. Type: Head straight with long and fairly thick neck. Inclined to favour the Pekin in shape of body with broad deep breast. Strong legs set well back. Carriage almost upright. Weights: Drake 4.10 kg (9 lb) max.; duck 3.20 kg (7 lb). Colour: Black and white. Legs and webs dark.

Geese

The first poultry show of 1845 classified Common Geese, Asiatic or Knob Geese, and 'Any other variety'. The first Book of Standards described the Toulouse and Embden. Peculiarly enough, these two breeds monopolized our standards up to recent times, being the chief ones exhibited regularly at shows. At times other breeds have been exhibited, and now the standards have been extended. The Greylag is said to be the ancestor of all our domestic geese, and the common goose of this country undoubtedly was the English Grey, although a white variety existed, and the Grey Back (Saddleback) may have come from an intercross.

The British Waterfowl Association classifies the following as Ornamental Geese: Canada, Egyptian and all breeds of British or foreign wild geese.

African

Origin: Africa

The African goose is the large relative of the Chinese, both having been developed from the wild swan goose which of course both resemble. It is a heavyweight compared to the Toulouse and Embden.

General characteristics: male and female

Carriage: Reasonably upright.

Type: Body large, long, carried moderately upright. Stern round, full, free from bagginess. Back broad, moderately long, flat. Breast full, well rounded, carried moderately upright, without keel. Wings large, strong, smoothly folded against sides. Tail well elevated.

Head: Broad, deep, large. Dewlap large, heavy, smooth; lower edges regularly curved and extending from lower mandible to below juncture of neck and throat. Bill rather large, stout at base. Knob large, broad as the head, protruding slightly forward from front of skull at upper mandible. Eyes large.

Neck: Long, nicely arched; throat with well-developed dewlap.

Legs and feet: Lower thighs short, stout. Shanks of medium length. Straight toes connected by web.

African

African gander

Colour

Plumage of both sexes: Head light brown. Neck very light ashy brown with distinct broad, dark brown stripe down centre of the back of neck and extending its entire length. Front of neck under mandible very light ashy brown, gradually getting lighter in colour until past the dewlap where it is almost cream in colour, then gradually deepening in colour as it approaches the breast. Breast very light ashy brown shading to a lighter colour under the body. Body a lighter shade than the breast, gradually getting lighter as it approaches the fluff, which is so light as to approach white; sides of body ashy brown, each feather edged with a lighter shade. Lower thighs: upper part similar to sides of body, ashy brown edged with a lighter shade, lower part similar in colour to underpart of body. Back ashy brown. Wing bow ashy brown, slightly edged with a lighter shade, coverts ashy brown, distinctly edged with a lighter shade; primaries dark slate; primary coverts light slate; secondaries dark slate, distinctly edged with a lighter shade approaching white. Tail ashy brown heavily edged with a shade approaching white, tail coverts white.

In both sexes: Bill and knob black. Eyes dark brown. Legs and webs dark orange.

Weights

Adult gander 9.10 kg (20 lb); young gander 7.25 kg (16 lb)
Adult goose 8.15 kg (18 lb); young goose 6.35 kg (14 lb)

White African

Scale of points

Size	20
Colour and markings	10
Condition	10
Legs and feet	5
Breast	10
Head and throat	15
Neck	5
General carriage	15
Tail and paunch	10
	100

White African

White African geese are identical to grey African geese in shape requirements, but the colour details are similar to those specified for white Chinese, including disqualifications.

White African Goose and gander

American Buff gander

American Buff

Origin: America

The American buff goose is a medium-sized breed with a stance like the Embden, smooth breasted and dual-lobed.

General characteristics: male and female

Carriage: Upright.

Type: Body moderately long, broad, plump. Back medium in length, broad and smooth, and slightly convex from shoulder to tail. Breast broad, deep and full. Wings medium in size and smoothly folded close to body. Tail medium in length with broad, stiff feathers.

Head: Medium in size, moderately broad, oval, strong. Bill of medium length, stout and tapering evenly to well-rounded end. Eyes large and full.

Neck: Medium in length, slightly arched, rather upright and strong in appearance.

Brecon Buff

Legs and feet: Lower thighs medium length, well fleshed. Shanks stout, straight, moderately long. Toes straight and well webbed.

Colour

Plumage of both sexes: Head soft, dull buff. Neck soft, dull buff shading lighter in front as it approaches the breast, which is very light buff. Back dark buff. Body and fluff light buff growing lighter until almost white in abdomen, continuing back to the tail. Sides lighter buff, growing darker over thighs dull light buff. Wing bow and coverts soft, dull buff with narrow edging of very light greyish buff; primaries dull buff, secondaries darker buff with slight edgings of lighter buff. Tail buff and white, the ends of feathers edged with very pale buff.

In both sexes: Bill light orange. Eyes dark hazel. Legs and webs orange.

Weights

Adult gander 8.15 kg (18 lb); young gander 7.25 kg (16 lb)
Adult goose 7.25 kg (16 lb); young goose 6.35 kg (14 lb)

Scale of points

Size	10
Colour and markings	40
Condition	10
Legs and feet	5
Breast	10
Head and throat	5
Neck	5
General carriage	10
Tail and paunch	5
	100

Brecon Buff

Origin: Great Britain

At different times attempts have been made to create a buff goose, and to Wales goes the credit of originating the Brecon Buff, founded on stock from Breconshire hill farms, the breed being recognized officially in 1934. Hardy, light in bone, with maximum flesh, and an active breed, it is also more prolific than the ultra-heavy breeds.

Brecon Buff

Brecon Buff gander

General characteristics: male and female

Carriage: Upright and alert, indicative of activity.

Type: Body broad, well rounded and compact. Breast round, only slight indication of keel. Wings large and strong.

Head: Small and neat, no sign of coarseness. Eyes bright.

Neck: Long and thin, the throat showing no gullet.

Legs and feet: Legs fairly short. Strong shanks. Straight toes connected by web.

Plumage: Hard and tight.

Colour

Plumage of both sexes: A deep shade of buff throughout with markings similar to those of the Toulouse. Ganders are usually not as deep coloured as geese.

In both sexes: Bill pink. Eyes dark brown. Legs and webs pink or orange.

Buff Back

Weights

Gander 8.60 kg (19 lb) max.
Goose 7.25 kg (16 lb) max.

Scale of points

Type (breast 20, head and neck 15, general carriage 10)	45
Size	15
Legs and feet	5
Colour and markings	25
Condition	10
	100

Buff Back

Origin: Europe

General characteristics: male and female

Carriage: Nearly horizontal.

Type: Body moderately long, plump, deep and meaty with no evidence of a keel. Back slightly convex and approximately 60% longer than broad. Paunch moderately broad, deep and double-lobed. Wings rather long with the tips crossing over the tail coverts. Carried high and smoothly folded. Tail somewhat short, closely folded and carried nearly level.

Head: Fairly broad and somewhat refined with a nearly flat crown. Bill medium in length, nearly straight and stout. Eyes large and rather prominent.

Neck: Medium length, moderately stout and carried rather upright with little or no indication of an arch.

Legs and feet: Lower thighs medium length, plump and nearly concealed by ample thigh coverts. Shanks moderately long and rather refined, yet sturdy.

Colour

Plumage of both sexes: Head buff. Neck buff on the upper part and white on the lower. Back buff with each feather edged with near white from a point above; the scapulars are the same colour. The general appearance of this buff suggests a heart-shape. Breast, body, wings and tail white, except for a broad band of buff edged with near-white beginning under the secondaries (just over the shanks) and extending under the abdomen to the opposite wing. Remainder of plumage white, except for large thigh coverts, which are buff edged with white.

In both sexes: Bill orange. Eyes blue. Legs and webs reddish orange.

Buff Back

Buff Back Gander

Weights

Adult gander 7.70 kg (17 lb); young gander 6.80 kg (15 lb)
Adult goose   6.80 kg (15 lb); young goose 5.90 kg (13 lb)

Scale of points

Size	10
Colour and markings	40
Condition	10
Legs and feet	5
Breast	10
Head and throat	5
Neck	5
General carriage	10
Tail and paunch	5
	100

Chinese

Origin: Asia

A prolific breed of the smaller-bodied type of goose is the Chinese, which has of recent years become popular on the show-bench. Its popularity has grown and it long since became standardized in this country. It has at times been known as the Knob, Asiatic and Hongkong goose.

General characteristics: male and female

Carriage: Upright, compact and active.

Type: Body compact and plump. Back reasonably short, broad, flat, and sloping, to give a characteristic upright carriage. Breast well rounded and plump, and carried high. Wings large, strong and high-up, carried closely. Stern well rounded, and well-developed paunch. Tail close and carried well out.

Head: Medium for size, and proportionate. Bill stout at base, symmetrical, medium for size. Knob, large rounded and prominent (smaller in the goose than the gander). Eyes bold.

Neck: Long, swan-like, carried upright, but with graceful arch and refined (longer in the gander than the goose).

Legs and feet: Legs reasonably short. Shanks strong, medium for length. Toes straight, well spread and webbed.

Plumage: This should be reasonably tight.

Colour

The white
Plumage of both sexes: Pure white.

In both sexes: Bill and knob orange. Eyes blue. Legs and webs orange-yellow.

The brown-grey
Plumage of both sexes: Head dark russet brown, with face fawn up to demarcation line above the eyes. Face band a definite white band or line from top of head to as far down the face as possible. Neck fawn with prominent dark russet brown stripe down middle of back of neck and for its entire length. Back russet brown. Breast greyish fawn from under mandible well down to body where it becomes lighter. Thighs at side of breast russet brown, each feather edged with a lighter shade of greyish fawn, approaching white. Wing bow and coverts medium russet brown, each feather laced with a lighter greyish fawn edging, approaching white; flights russet brown. Stern, paunch and tail a lighter shade of greyish fawn, approaching white, the tail having a broad band of russet brown across with the light edging.

In both sexes: Bill black or dark slate. Knob dark slate. Eyes brown. Legs and web orange.

Chinese

Chinese
White gander
Grey gander

Embden

Weights

Gander 4.55–5.45 kg (10–12 lb)
Goose 3.60–4.55 kg (8–10 lb)

Scale of points

Type and carriage	30
Head points and neck	25
Colour	15
Size	10
Condition	10
Legs and feet	10
	100

Disqualifications

Absence of knob and heavy gullet.

Embden

Origin: North Europe

As the Embden breed was also known originally as the Bremen one associates it with Germany although stock reached us from Holland. In Germany and North Holland, no doubt they crossed the Italian white with their native whites, creating the Embden. When stock did reach this country our breeders crossed the birds with our own English whites, and by careful selection increased the body weight and quantity of meat, while standardizing the breed for characteristics.

General characteristics: male and female

Carriage: Upright and defiant.

Type: Body broad, thick and well rounded. Back long and straight. Breast round with very little, if any, indication of keel. Shoulders and stern broad. Paunch deep. Wings large and strong. Tail close and carried well out.

Head: Long and straight. Bill fairly short and stout at the base. Eyes bold.

Neck: Long and swan-like, the throat uniform with the under mandible and neck, i.e. without a gullet.

Legs and feet: Legs fairly short. Shanks large and strong. Toes straight and connected by web.

Plumage: Hard and tight.

Embden

Embden gander

Colour

Plumage of both sexes: Pure glossy white.
In both sexes: Bill orange. Eyes light blue. Legs and webs bright orange.

Weights

Gander 13.60–15.40 kg (30–34 lb)
Goose 9.10–10.00 kg (20–22 lb)

Scale of points

Type (breast 20, head 12, general carriage 12, neck 10)	54
Size	20
Colour	10
Condition	10
Legs and feet	6
	100

Serious defects

Plumage other than white. Any deformity.

Grey Back

Origin: Germany

The Grey Back goose, known in America as the Pomeranian goose, is unique amongst domestic geese for its single-lobed paunch and striking colour pattern. It is similar in size and stance to the American Buff.

General characteristics: male and female

Carriage: Nearly horizontal.

Type: Body moderately long, plump, deep and meaty with no evidence of a keel. Back slightly convex and approximately 60% longer than broad. Paunch moderately broad, deep and single-lobed. Wings rather long with the tips crossing over the tail coverts. Carried high and smoothly folded. Tail somewhat short, closely folded and carried nearly level.

Head: Fairly broad and somewhat refined with a nearly flat crown. Bill medium in length, nearly straight and stout. Eyes large and rather prominent.

Neck: Medium length, moderately stout and carried rather upright with little or no indication of an arch.

Legs and feet: Thighs medium length, plump and nearly concealed by ample thigh coverts. Shanks moderately long and rather refined, yet sturdy.

Colour

Plumage of both sexes: Head dark grey. Neck dark grey on the upper part and white on the lower. Back dark grey with each feather edged with near white from a point above, the scapulars being the same colour. The general appearance of this grey area suggests a heart-shape. Breast, body, wings and tail white, except for a broad band of blue-grey edged with near-white beginning under the secondaries (just over the shanks) and extending under the abdomen to the opposite wing. Remainder of plumage white, except for large thigh coverts which are grey, edged with white.

In both sexes: Bill reddish pink or a deep flesh colour. Eyes blue. Legs and webs reddish pink.

Weights

Adult gander 7.70 kg (17 lb); young gander 6.80 kg (15 lb)
Adult goose 6.80 (15 lb); young goose 5.90 (13 lb)

Scale of points

Size	10
Colour and markings	40
Condition	10
Legs and feet	5
Breast	10

(continued)

(**Scale of points** – *continued*)
Head and throat	5
Neck	5
General carriage	10
Tail and paunch	5
	100

Pilgrim

Origin: Great Britain

The Pilgrim goose was developed in the West of England and is unique because of its sex-linked plumage colour.

General characteristics: male and female

Carriage: Above the horizontal, but not upright.

Type: Body moderately long, plump and meaty; keel permissible in goose. Adult abdomen deep, square and well balanced, free from bagginess. Back moderately broad, uniform in width, flat and straight. Breast round, full, deep. Wings strong, well developed, neatly carried to body. Tail medium in length, closely folded, carried nearly level.

Head: Medium in size, oval, trim. Bill medium in length, straight, stout, smoothly attached. Eyes moderately large.

Neck: Medium in length, moderately stout, slightly arched.

Legs and feet: Lower thighs medium in length, well fleshed. Shanks moderately short and stout. Toes, strong, straight, and well webbed.

Plumage: Hard, tight and glossy.

Colour

Gander's plumage: Pure white. Some hidden grey permissible in back plumage, and in wings, back and tail of young ganders.

Goose's plumage: Head light grey and white, white predominating in adult geese. Neck light grey, upper portion mixed with white in mature specimens. Back light ashy grey, laced with lighter grey. Breast very light ashy grey, gradually getting lighter as it approaches fluff which is so light as to approach white. Sides of body soft, ashy grey, each feather edged with lighter shade. Wing bow and coverts light ashy grey, edged with lighter grey; primaries medium grey, coverts light grey, secondaries medium grey with lighter shade approaching white. Tail ashy grey, heavily edged with a lighter grey approaching white.

In both sexes: Bill orange. Eyes bluish grey in gander, hazel brown in goose. Legs and webs orange.

Roman

Weights

Adult gander 6.35 kg (14 lb); young gander 5.45 kg (12 lb)
Adult goose 5.90 kg (13 lb); young goose 4.55 kg (10 lb)

Scale of points

Symmetry	4
Weight	4
Condition	10
Bill	6
Eyes	4
Head	10
Neck	8
Back	10
Tail	4
Wings	10
Breast	14
Body	13
Legs and feet	3
	100

Roman

Origin: Mediterranean

Another of the smaller type of goose, the Roman was introduced into England from Italy about 1903, and there were other importations at later dates. Earliest arrivals often were marked with grey on the back, but were eliminated by selective matings for the pure white.

General characteristics: male and female

Carriage: Active, alert, graceful, docile rather than defiant, with horizontal outline (particularly in goose).

Type: Compact and plump, deep and broad, but reasonably long and well balanced. Back wide, flat and with gentle slope that is more pronounced in gander than goose. Breast full, well rounded, somewhat low and without prominence of keel. Wings large, long, strong, high-up and well tucked up to tail line. Stern well rounded off, paunch not too pronounced. Tail close, long and carried well out.

Head: Neat and well rounded (especially in goose) symmetrical and refined. Face deep (particularly in goose). Bill short and not coarse. Eyes bold, well up in skull.

Neck: Upright, medium length, refined (particularly in goose) and without gullet.

Legs and feet: Legs short, light boned, well apart. Toes straight and connected by web.

Plumage: Sleek, short, tight and with glossy feathering.

Roman goose

Handling: High proportion of meat to bone and offal. To be emphasized in judging.

Colour

Plumage of both sexes: Glossy white.
 In both sexes: Bill orange-pink. Eyes light blue. Legs and webs orange.

Weights

Gander 5.45–6.35 kg (12–14 lb)
Goose 4.55–5.45 kg (10–12 lb)

Scale of points

Type and carriage	20
Table qualities	20
Head and neck	15
Colour	15
Ideal size	10
Condition	10
Legs and feet	10
	100

Serious defects

Plumage other than white. Any deformity. Excessive weight or bone. Coarseness.

Sebastopol

Origin: Eastern Europe

The Sebastopol is one of the most unusual of the breeds of domestic geese. The long frizzled or spiralled feathers on the breast and the loose fluffed plumage make the Sebastopol a unique and attractive breed. Though grey and buff crosses with poor feather curl have appeared, the white variety is the only true Sebastopol at present.

The breed is primarily one for the exhibitor but the Sebastopol is a medium weight breed, a moderate egg-layer and a fast grower and thus has merit as a utility goose. They are a quiet, friendly breed and are thus a good domestic breed for the small scale breeder.

General characteristics: male and female

Carriage: Horizontal.

Type: Body appears round because of the full feathering. Back of medium length but appears short because the long feathers give body a rounded ball appearance. Breast full and deep, lacking keel. Wing feathers long, well curled and flexible. They should be so soft as to make the bird incapable of flight and also entirely devoid of stiff shafts. Tail composed of long, well-curled feathers.

Head: Appears large in proportion to body. Bill of medium length. Eyes large and prominent.

Neck: Medium length and carried rather upright instead of forward.

Legs and feet: Lower thighs short but stout, each covered with curled feathers. Shanks short and stout.

Plumage: Only feathers of head and upper neck smooth. Feathers on lower neck, breast and remainder of body profusely curled. Feathers of wings and back should be long (the longer the better), well curled and free from stiff shafts. Specimens of good stock and in good condition should display back and wing feathers that should almost touch the ground.

Colour

Plumage of both sexes: Pure white in colour, though traces of grey in young specimens allowed and will not result in reduced points when judged.

In both sexes: Bill and webs orange, the webs having deeper shade. Eyes bright blue.

Weights

Adult gander 6.35 kg (14 lb); young gander 5.45 kg (12 lb)
Adult goose 5.45 kg (12 lb); young goose 4.55 kg (10 lb)

Sebastopol

Sebastapol
Gander and goose

Scale of points

Size	10
Condition and vigour	10
Shape (head, bill and eyes)	10
Neck	2
Back	4
Breast	4
Paunch	4
Tail	2
Legs	2
Feet	2
Feathering	20
Style and carriage	10
Colour	20
	100

Toulouse

Origin: France

France originated the Toulouse and developed it for table purposes. Stock was sent over to England and our breeders crossed the birds with our own English Greys, and developed the breed for body weight and quantity of flesh, as well as standard characteristics of plumage colour, markings and type.

General characteristics: male and female

Carriage: Thick set and somewhat horizontal, but not as upright in front as the Embden.

Type: Body long, broad and deep. Back slightly curved from the neck to the tail. Breast prominent, deep and full, the keel straight from stem to paunch, increasing in width to the stern and forming a straight underline. Shoulders broad. Wings large and strong. Tail somewhat short, carried high and well spread. Paunch and stern heavy and wide, with a full rising sweep to the tail.

Head: Strong and massive. Bill strong, fairly short and well set in a uniform sweep, or nearly so, from the point of the bill to the back of the skull. Eyes full.

Neck: Long and thick, the throat well gulleted.

Legs and feet: Legs short. Shanks stout and strong boned. Straight toes connected by web.

Plumage: Full and somewhat soft.

Colour

Plumage of both sexes: Neck dark grey. Breast and keel rather light grey, shading dark to thighs. Back, wings and thighs dark steel grey, each feather laced with an almost white edging, the flights without white. Stern, paunch and tail white, the tail with broad band of grey across the centre.

In both sexes: Bill, legs and webs orange. Eyes dark brown or hazel.

Weights

Gander 12.70–13.60 kg (28–30 lb)
Goose   9.10–10.00 kg (20–22 lb)

Scale of points

Type (head and throat 15, breast and keel 10, tail, stern and paunch 10, neck 5, general carriage 15)	55
Size	20

(continued)

Toulouse goose

(**Scale of points** – *continued*)
Legs and feet	5
Colour and markings	10
Condition	10
	100

Serious defects

Patches of black or white among the grey plumage. Slipped or cut wings. Any deformity.

Other breeds

Among other breeds of geese occasionally exhibited in this country is the *English*, with two varieties, grey and white, very similar in colour to the Toulouse and Embden but generally half the size and weight of those breeds.

Standard for eggs

THE POULTRY CLUB has authorized the following standard and scale of points for judging eggs.

External

Shape: Showing ample breadth, good dome, with greater length than width, the top to be much roomier than the bottom and more curved. The bottom is more pointed in the hen egg than that of the pullet, but it should not be too pointed, and a circular, or even narrow shape is undesirable.

Size: Mere size is not a deciding point, without texture of shell combined. A pullet's normal egg when the bird starts to lay is 49.6 g (1¾ oz) and increases quickly to 56.7 g (2 oz), exceeding that after several months of production. There is another increase in the hen egg after the moult. Bantam eggs should be judged to the same standard as for large fowl eggs with the exception of size which should ideally not exceed 42.5 g (1½ oz). Eggs weighing in excess of this weight should be discouraged or passed.

Shell texture: Smooth, free from lines or bulges, evenly limed, smooth at each end, without roughness, porous parts or lime pimples.

Colour: Brown, tinted, white, cream, mottled, blue, green, olive, plum, etc.

Freshness, bloom and appearance: Shells to be clean, without dull or stale appearance; airspace to be small, as befits a new-laid egg, contents on candling to be free of blood or meat spots. Eggs may be washed in preparation.

Uniformity: Eggs forming a plate or exhibit, to be uniform in shape, shell grade, shell texture, size and colour.

Internal

Yolk: Rich bright golden yellow, free from blood streaks or spots. Well rounded, smooth on surface, and well raised from albumen. One uniform shade. Blastoderm or germ spot not discoloured.

Albumen: Preferably white in colour, of dense substance, particularly around the yolk, which it raises. Outline of albumen to be seen. Free of blood spots.

Chalazæ: Each chalaza to resemble a thick cord of white albumen at each end of yolk, keeping it to centre of first or thickest albumen. Free of blood spots. Other layers of albumen less dense.

Standard for Eggs

(*a*) **Large fowl: single white**

(*b*) **Large fowl: single brown**

(*c*) **Bantam: six brown**

(*d*) **Large fowl: six white**

(*e*) **Bantam: three distinct colours**

(*f*) **Bantam: three cream**

Standard for Eggs

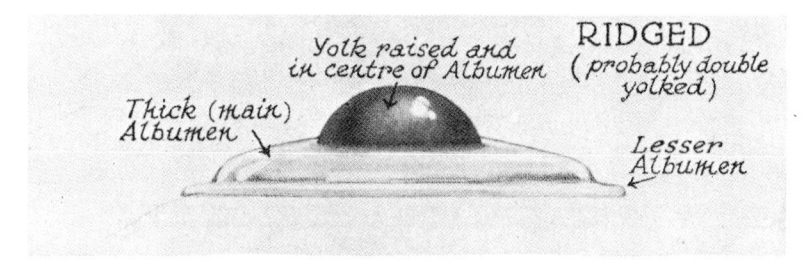

Illustration labels: Thick (main) Albumen; Yolk raised and in centre of Albumen; RIDGED (probably double yolked); Lesser Albumen

Airspace: Very small, as befits a new-laid egg, the membrane adhering to shell.

Freshness: Indicated by small airspace, and unwrinkled top surface of yolk, and its height. Stale yolks flop at edges.

Scale of points

External

Shape	25
Size	15
Shell texture	20
Colour	20
Freshness, bloom, and appearance	20
	100*

Internal

Yolk	35
Albumen	35
Chalazæ	10
Airspace	10
Freshness	10
	100

* May be maximum for each egg, or for a plate of eggs, whatever the number. Add 5 points more for each egg for 'matching and uniformity'.

Defects (for which eggs may be passed)

More than one yolk. Staleness. Polished or overprepared shells. Defective contents even when judged for externals. Addition of colouring to shells. Artificial polish or colouring would amount to disqualification and report to Poultry Club.

Glossary

Abdomen: Underpart of body from keel to vent.

A.O.C.: Any other colour.

A.O.V.: Any other variety.

Axial feather: Small feather between wing primaries and secondaries.

Back: Top of body from base of neck to beginning of tail.

Bands: see 'Pencilling'.

Bantam: Miniature fowl, formerly accepted as one-fifth the weight of the large breed it represented, but nowadays about one-fourth.

Barring: Alternate stripes of light and dark across a feather, most distinctly seen in the barred Plymouth Rock.

Bay: A reddish brown colour (see also 'Wing bay').

Beak: The two horny mandibles projecting from the front of the face.

Bean: A black spot or mark (generally raised) at the tip of the upper mandible of a duck's bill, seen in Cayugas and other breeds of waterfowl.

Beard: A bunch of feathers under the throat of some fowls, such as the Faverolles, Houdan and some varieties of Poland. A tuft of coarse hair growing from the breast of an adult turkey male, also known as the 'tassel'.

Beetle brows: Heavy overhanging eyebrows, best seen in the Malay.

Bill: A duck's beak.

Blade: Rear part of a single comb.

Blocky: Heavy and square in build.

Booted: Feathers projecting from the shanks and toes, as in the Brahma, Cochin, and Booted bantam.

Bow legged: Greater distance between legs at the hocks than at knees and feet.

Brassiness: Yellowish foul colouring on plumage, usually on back and wing.

Glossary

Breast: Front of a fowl's body from point of keel bone to base of the neck. In dead birds, flesh on the keel bone.

Breed: A group of birds answering truly to the type, carriage and characteristics distinctive of the breed name they take. There may be varieties within a breed, distinguished by differences of colour and markings.

Cap: A comb; also the backpart of a fowl's skull.

Cape: Feathers under and at base of neck hackle, between the shoulders.

Capon: Strictly speaking a castrated male fowl, but term is also used to describe one treated chemically.

Carriage: The bearing, attitude, or style of a bird, especially when walking.

Caruncles: Fleshy protuberances on head and wattles of turkeys and Muscovy ducks.

Chicken: A term employed by the Poultry Club to describe a bird of the current season's breeding.

Cinnamon: A dark reddish buff colour.

Cloudy: Indistinct (see 'Mossy').

Cobby: Short, round or compact in build.

Cock: A male bird after the first adult moult.

Cockerel: A male bird of the current year's breeding.

Cockerel-breeder: A term applied to birds, either male or female, selected to produce good standard bred cockerels.

Collar: A white mark almost enhancing the neck of the Rouen drake, also known as the 'ring'.

Comb: Fleshy protuberance on top of a fowl's head, varying considerably in type and size and including cushion (Silkie), horn or V-shaped (La Flèche and Sultan), leaf or shell (Houdan), pea or triple (Brahma), rose (Hamburgh, Wyandotte, etc.), single (Cochin, Leghorn, etc.), cap (Sicilian Buttercup), strawberry or walnut (Malay), and raspberry (Orloff).

Concave sweep: Hollow curve from shoulders to part way up the tail.

Condition: State of a bird's health, brightness of comb and face and freshness of plumage.

Coverts: Covering feathers on tail and wings.

Cow hocks: Weakness at hocks (see 'Knock-kneed').

Types of comb: 1 Rose, leader following line of neck. 2 Triple or pea. 3 Rose, short leader. 4 Walnut. 5 Cap. 6 Mulberry. 7 Medium single. 8 Large single. 9 Cup. 10 Rose with long leader. 11 Leaf. 12 Horn. 13 Small single. 14 Folded single. 15 Semi-erect single.

Glossary

Glossary

Crescent: Shaped like the first or last quarter of the moon.

Crest: A crown or tuft of feathers on the head; known also as 'top knot' and in Old English Game as the 'tassel'.

Crow head: Head and beak narrow and shallow, like a crow.

Cuckoo barring: Irregular barring where the two colours are somewhat indistinct and run into each other, as in the North Holland Blue, cuckoo Leghorn and Marans.

Cup comb: A comb somewhat resembling a tea-cup with the edges spiked as in the Sicilian Buttercup.

Cushion: A mass of feathers over the back of a female covering the root of her tail, and most prominently developed in the Cochin.

Cushion comb: An almost circular cushion of flesh, with a number of small prominences over it, and having a slight furrow transversely across the middle, as in the Silkie.

Daw eyed: Having pearl-coloured eyes.

Deaf-ears: see 'Ear-lobes'.

Dewlap: The gullet (so called), seen to the best advantage in adult Toulouse geese. Loose pouch of skin on throat under the beak.

Diamond: The wing bay. A term commonly used among Game fanciers.

Dished bill: Depression or hollow in the upper line of the bill of a duck or drake.

Dished lobe: Lobe that is hollow in the centre.

Double comb: see 'Rose comb'.

Double laced: Two lacings of black as on an Indian Game female's feather. First there is the outer black lacing round the edge of the feather and next the inner or 'second' lacing (see 'Lacing').

Down: Initial hairy covering of baby chick, ducklings, etc. Also the fluffy part of the feather below the web and small tufts sometimes seen as faults on toes and legs of clean legged breeds (see 'Fluff').

Drake: Male duck.

Dubbing: Removal of comb, wattles, and ear-lobes, so as to leave a fowl's head smooth.

Duck: General term for certain species of waterfowl, and also used to describe the female.

Duck footed: Fowls having the rear toe lying close to the foot instead of spread out, thus resembling the foot of a duck.

Dusky: Yellow pigment shaded with black.

Glossary

Ear-lobes: Folds of skin hanging below the ears, sometimes called 'deaf-ears'. They vary in size, shape, and colour, the last named including purple-black, turquoise-blue, cream, red, and white.

Face: The skin in front of, behind, and around the eyes.

Feather legged: Characteristic of breeds such as the Brahma, Cochin, Faverolles, etc. May be sparsely feathered down to the outer toes, as in the Faverolles, or profusely feathered to the extremity of middle and outer toes as in the Brahma. Serious defect in clean legged breeds.

Flat shins: Shanks that are flat fronted instead of rounded.

Flight coverts: Small stiff feathers covering base of the primaries.

Flights: Primary feathers of the wings used in flying, but tucked out of sight when the bird is at rest.

Fluff: Soft downy feathers around the thighs, chiefly developed in birds of the Cochin type; the downy part of the feather (the undercolour) not seen as a rule until the bird is handled; also the hair-like growth sometimes found on the shanks and feet of clean legged fowls, and in this case a defect.

Footings: see 'Booted'.

Foxy: Rusty or reddish in colour (see also 'Rust').

Frizzled: Curled; each feather turning backwards so that it points towards the head of the bird.

Furnished: Feathered and adorned as an adult. A cockerel that has grown his full tail, hackles, comb, etc. is said to be 'furnished'.

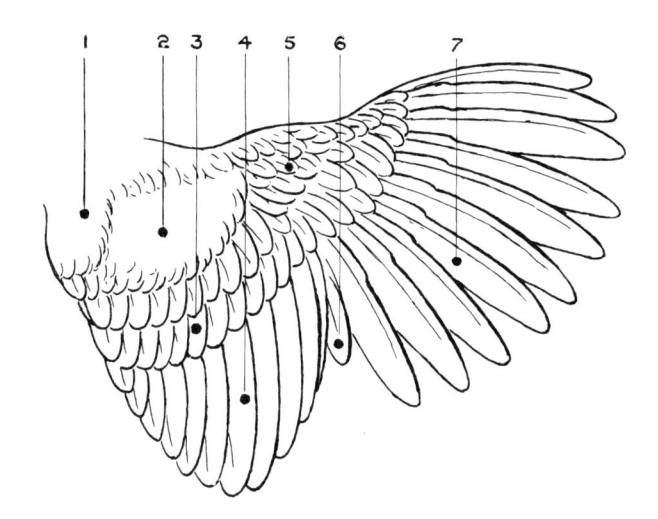

1 Shoulder butt. 2 and 5 Bow coverts. 3 Bar. 4 Secondaries. 6 Axial feather. 7 Primaries.

Glossary

Gander: The male of geese.

Gay: Excess white in markings of plumage.

Gypsy face: The skin of the face a dark purple or mulberry colour.

Goose: The female of geese.

Grizzled: Grey in the flights of an otherwise black bird.

Ground colour: Main colour of body plumage on which markings are applied.

Gullet: The loose part of the lower mandible; the dewlap of a goose. It appears on fowls, and is seen most distinctly perhaps on old Cochin hens, when it resembles a miniature beard of feathers.

Hackles: The neck feathers of a fowl and the saddle plumage of a male, consisting of long, narrow, pointed feathers.

Hangers: Feathers hanging from the posterior part of a male fowl – the lesser sickles and tail coverts known as tail hangers, and the saddle hackle as saddle hangers.

Hard feather: Close tight feathering as found on Game birds.

Head: Comprises skull, comb, face, eyes, beak, ear-lobes and wattles.

Hen: A female after the first adult moult.

Hen feathered: A male bird without sickles or pointed hackles (sometimes called a 'henny').

Hind toe: The fourth or back toe of a fowl.

Hock: Joint of the thigh with the shank, sometimes called the knee or the elbow.

Hollow comb: Depression in comb.

Hollow lobes: Depression in ear-lobes.

Horn comb: A comb said to resemble horns, but generally similar to the letter V, and seen to the best advantage on a matured Flèche or Sultan male. The comb starts just above the beak, and from it branch two spikes thick at the base and tapered at the end.

In-kneed: see 'Knock-kneed'.

Iris: Coloured portion of eye surrounding the pupil.

Keel: Blade of the breast bone; in ducks the dependent flesh and skin below it. In geese the loose pendent fold of skin suspended from the underpart of the bone.

Glossary

Keel bone: Breast bone or sternum.

Knob: Protuberance on upper mandible of certain brands of geese.

Knock-kneed: Hocks close together instead of well apart.

Lacing: A stripe or edging all round a feather, differing in colour from that of the ground; single in such breeds as the Andalusian, Wyandotte, and Sebright bantams, and double in Indian Game and other females. In the last case the inner lacing not as broad as the outer (see also 'Double lacing').

Leader: The single spike terminating the rose type of comb; also known as the 'spike'.

Leaf comb: A comb resembling the shape of a butterfly with its wings nearly wide open, and the body of the insect resting on the front of the fowl's head. It has also been referred to as resembling two escallop shells joined near the base, the join covered with a piece of coral. Seen to the best advantage on a Houdan male.

Leg: The shank or scaly part.

Leg feathers: Feathers projecting from the outer sides of the shanks of such breeds as the Brahma, Cochin, Faverolles, Langshan and Silkie.

Lesser sickles: see 'Sickles'.

Lobes: see 'Ear-lobes'.

Lopped comb: Falling over to one side of the head.

Lustre: see 'Sheen'.

Main tail feathers: see 'Tail feathers'.

Mandibles: Horny upper and lower parts of beak or bill.

Marking: The barring, lacing, pencilling, spangling, etc. of the plumage.

Mealy: Stippled with a lighter shade, as though dusted with meal, a defect in buff-coloured fowls.

Moons: Round spangles on tips of feathers.

Mossy: Confused or indistinct marking; smudging or peppering. A defect in most breeds.

Mottled: Marked with tips or spots of different colour.

Muff: Tufts of feathers on each side of the face and attached to the beard, seen in such breeds as the Faverolles, Houdan, and some varieties of Poland; also known as 'whiskers'.

360

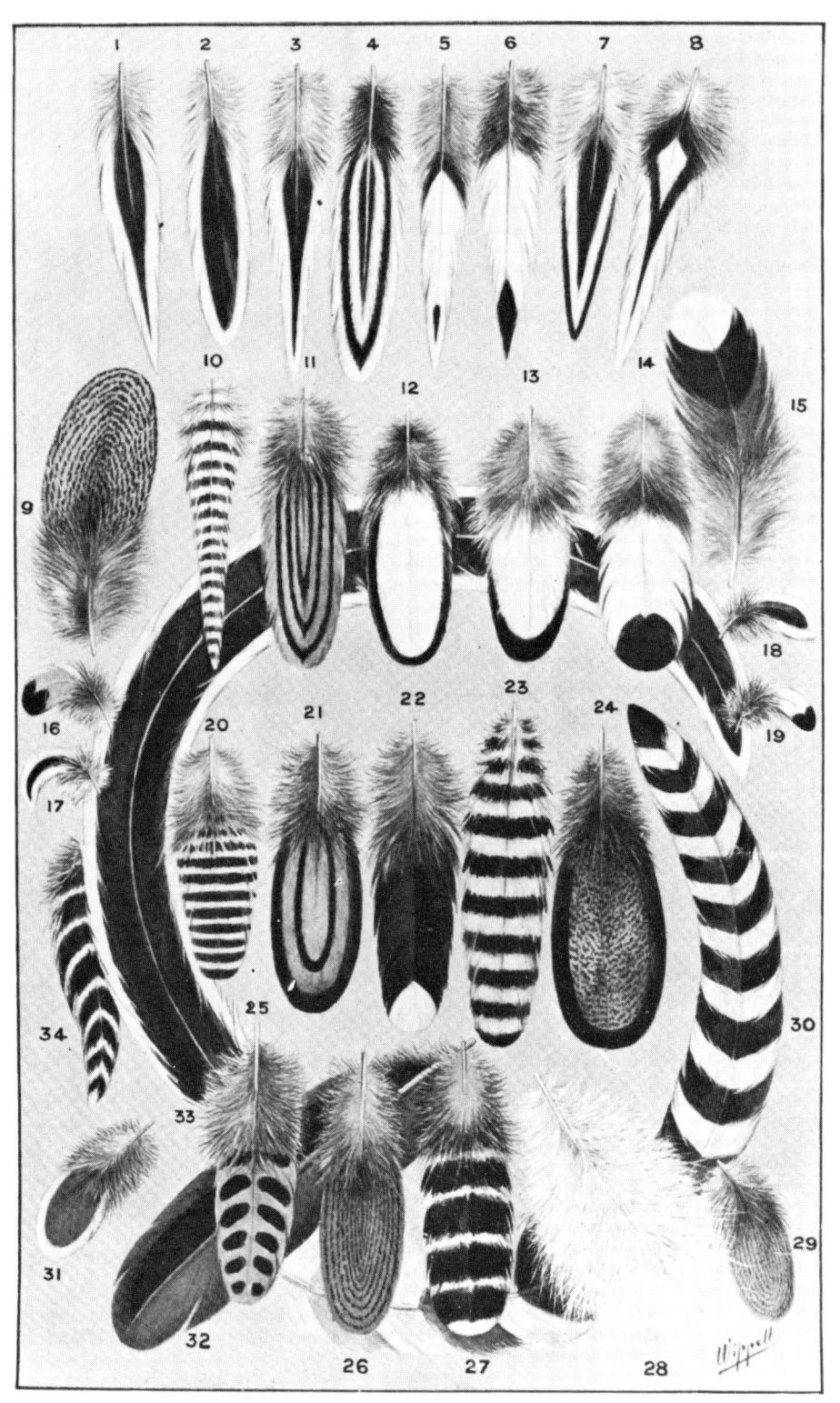

Glossary

Muffling: The beard and whiskers, i.e. the whole of the face feathering except the crest. In Old English Game the muffed variety has a thick muff or growth of feathers under the throat, differing in formation from that of the breeds named under 'Muff'.

Mulberry: see 'Gypsy face'.

Open barring: Where the bars on a feather are wide apart.

Open lacing: Narrow outer lacing, which gives the feather a larger open centre of ground colour.

Outer lacing: Lacing around the outer edge of a feather as opposed to 'inner' lacing.

Parti-coloured: Breed or variety having feathers of two or more colours, or shades of colour.

Pea comb: A triple comb, resembling three small single combs joined together at the base and rear, but distinctly divided, the middle one being the highest; best seen on the head of a well bred Brahma.

Pearl eyed: Eyes pearl coloured. Sometimes referred to as 'daw eyed'.

Pencilled spikes: The spikes of a single comb that are very long and narrow; little broader at the base than at the top; generally a defect.

Pencilling: Small markings or stripes on a feather, straight across in Hamburgh females (and often known as bands); or concentric in form, following the outline of the feather, as in the Brahma (dark), Cochin (partridge), Dorking (silver grey), and Wyandotte (partridge and silver pencilled) females, and fine stippled markings on females of Old English Game and brown Leghorns.

Peppering: The effect of sprinkling a darker colour over one of a lighter shade.

Primaries: see 'Flights'.

Feather markings: 1 Neck hackle, male (striped). 2 Neck hackle, female (laced). 3 Saddle hackle, male (striped). 4 Pencilled hackle (female). 5 Ticked hackle. 6 Tipped neck hackle, male, as in spangled Hamburgh. 7 Striped hackle, male. Shows outer fringing of colour – a fault. 8 Striped saddle hackle, male, showing open centre (desired only in pullet breeder). 9 Pencilled feather, cushion, female, as in silver grey Dorking and brown Leghorns. 10 Barred neck hackle (male). 11 Triple pencilled back (female). 12 Laced. 13 Faulty laced (i.e. horseshoed). 14 Spangled (moon-shaped). 15 Speckled. Irregular-shaped white tick shows three colours on feather. 16 Shoulder feather in spangled varieties. 17 Polish laced crest (pullet). 18 Polish crest, female. 19 Crescent marked. 20 Barred or finely pencilled as in Hamburgh. Bars and spaces same width. 21 Double laced. 22 Tipped, showing 'V'-shaped tip, as in Ancona. 23 Barred as in barred Rock, shows barring in undercolour. To finish with black bar. 24 Laced and ticked, as in dark Dorking. 25 Elongated spangle, as in Buttercup. 26 Finely pencilled, as in dark Brahma female. 27 Barred, as in Campine. Finishes with white end. Light bars a quarter to a third of the width of dark bars. 28 'Silkie' (no webbing). 29 Fine in pencilling, as in black marks of black-red, and duckwing Game. 30 Barred Rock sickle. 31 Buff laced. 32 Wing marking on flight feather. 33 Laced sickle. 34 Saddle hackle mackerel marked (Campine cockerel).

Glossary

Primary coverts: see 'Flight coverts'.

Pullet: A female fowl of the current season's breeding.

Pullet-breeder: A term applied to birds, either male or female, selected to produce good standard-bred pullets.

Pupil: Black centre of eye.

Quill: Hollow stem of feathers attaching them to the body.

Raspberry comb: A comb somewhat resembling a raspberry cut through its axis (lengthwise) and covered with small protuberances.

Reachy: Tall and upright carriage and 'lift' as in Modern Game.

Ring: see 'Collar'.

Roach back: Humped back.

Rose comb: A broad comb, nearly flat on top, covered with small regular points or 'work', and finishing with a spike or leader. It varies in length, width, and carriage according to breed.

Rust: A patch of red-brown colour on the wings of females of some breeds, chiefly those of the black-red colour; brown or red marking in black fluff or breast feathers; known also as 'foxiness' in females.

Saddle: The posterior part of the back, reaching to the tail of the male, and corresponding to the cushion in a female.

Saddle hackle: see 'Hackles'.

Sandiness: Giving the appearance of having been sprinkled with sand.

Sappiness: A yellow tinge in plumage.

Secondaries: The quill feathers of the wings which are visible when the wings are closed.

Self colour: A uniform colour, unmixed with any other.

Serrations: 'Saw tooth' sections of a single comb.

Shaft: The stem or quill part of the feather.

Leg types: 1 Clean legged, flat side (Leghorns). 2 Clean legged round shanks (Game). 3 Heavy feather legged, and feathered toes, i.e. foot feather. 4 Feather legged, no feathers middle toe (Croad Langshan). 5 Short round shanks (Indian Game). 6 Five toed (Dorking). 7 Slightly feathered shanks (Modern Langshan). 8 Feather legged and vulture hocked. 9 Thin round shanks (Modern Game). 10 Mottled shanks (Ancona). 11 Mottled and five toed (Houdan). 12 Feather legged and five toed (Faverolles).

Glossary

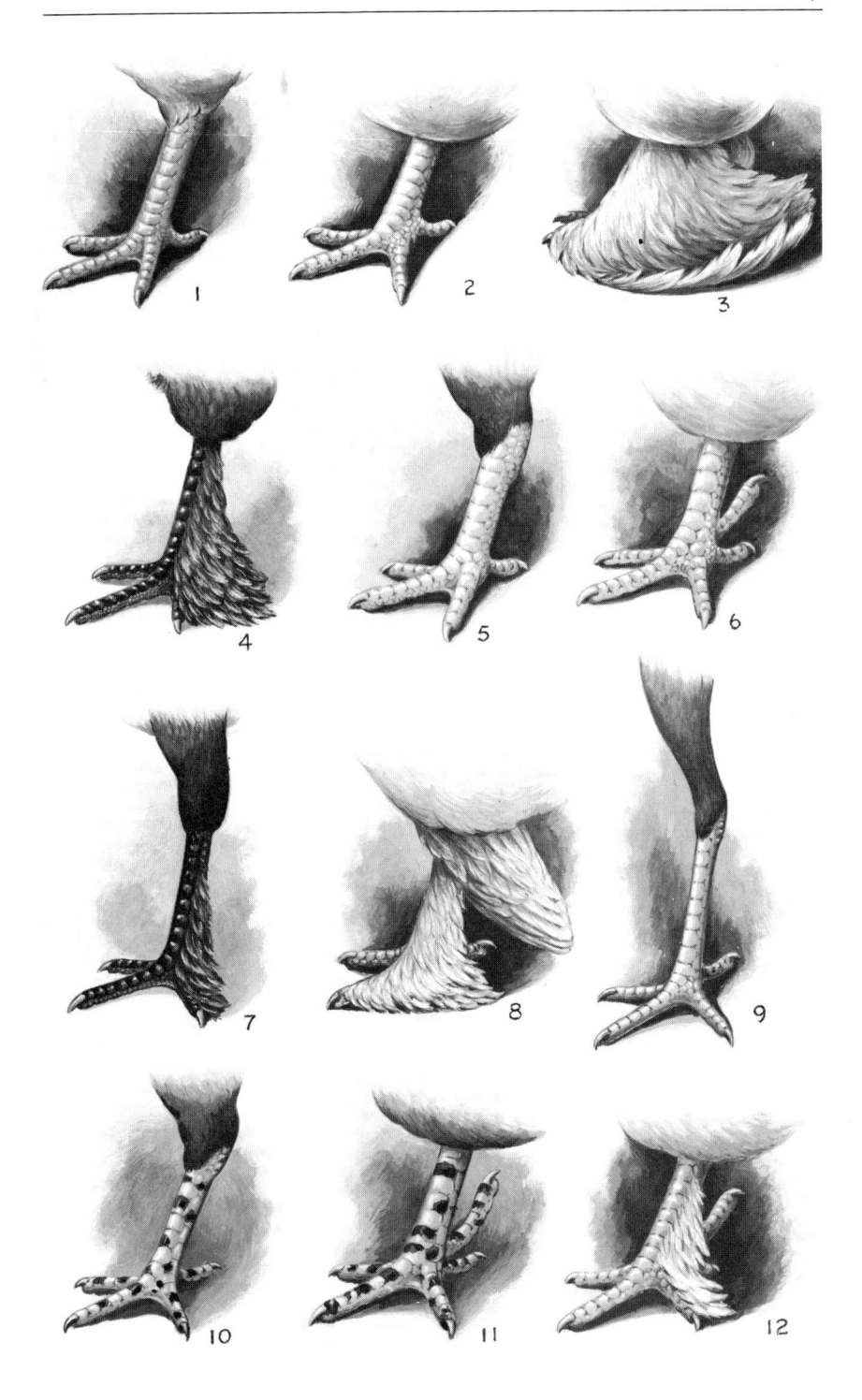

Glossary

Shafty: Lighter coloured on the stem than on the webbing; a desirable marking in dark Dorking females and Welsummers. Generally a defect in other breeds.

Shank: see 'Leg'.

Shank feathering: see 'Feather legged'.

Sheen: Bright surface gloss on black plumage. In other colours usually described as lustre.

Shell comb: see 'Leaf comb'.

Shoulder: The upper part of the wing nearest the neck feather. Prominent in Game breeds where it is often called the shoulder butt (see also 'Wing butt').

Sickles: The long curved feathers of a male's tail, usually applied to the top pair only (the others often being called the 'lesser' sickles), but sometimes used for the tail coverts.

Side sprig: An extra spike growing out of the side of a single comb.

Single comb: A comb which, when viewed from the front, is narrow, and having spikes in line behind each other. It consists of a blade surmounted by spikes, the lower (solid) portion being the blade, and the spaces between the spikes the serrations. It differs in size, shape, and number of serrations according to breeds.

Slipped wing: A wing in which the primary flight feathers hang below the secondaries when the wing is closed. This condition is often allied with split wing, in which primaries and secondaries show a very distinct segregation in many breeds of bantams.

Smoky undercolour: Defective grey pigment in the undercolour of a bird.

Smut: Dark or smutty colour where undesirable, such as in undercolour.

Soft feathered: Applied to breeds other than the hard feather group of Indian and Jubilee Game, Old English Game, Asil, Malay, and Modern Game.

Sootiness: Grey or smokiness creeping in where it is not wanted, usually in undercolour.

Spangling: The marking produced by a spot of colour at end of each feather differing from that of the ground colour. When applied to a laced breed, as the Poland, it means broader lacing at the tip of each feather. The spangle of circular form is the more correct, since, when of crescent or horseshoe shape, it favours the laced character.

Spike: The rear leader on a rose comb.

Splashed feather: A contrasting colour irregularly splashed on a feather.

Split comb: The rear blade of a single comb is split or divided.

Split crest: Divided crest that falls over on both sides.

Glossary

Split tail: Decided gap in middle of tail at base.

Split wing: see 'Slipped wing'.

Sprig: see 'Side sprig'.

Spur: A projection of horny substance on the shanks of males, and sometimes on females.

Squirrel tail: A tail, any part of which projects in front of a perpendicular line over the back; a tail that bends sharply over the back and touches, or almost touches, the head, like that of a squirrel.

Strain: A family of birds from any breed or variety carefully bred over a number of years.

Strawberry comb: A comb somewhat resembling half a strawberry, with the round part of the fruit uppermost; known also as the walnut comb.

Striping: The very important markings down the middle of hackle feathers, particularly in males of the partridge variety.

Stub: Short, partly grown feather.

Sub-variety: see 'Variety'.

Surface colour: That portion of the feathers exposed to view.

Sword feathered: Having sickles only slightly curved, or scimitar shaped, as in Japanese bantams.

Symmetry: Perfection of outline, proportion; harmony of all parts.

Tail coverts: see 'Coverts'.

Tail feathers: Straight and stiff feathers of the tail only. The top pair are sometimes slightly curved, but they are generally straight or nearly so. In the male fowl, main tail feathers are contained inside the sickles and coverts.

Tassel: see 'Crest' and 'Beard'.

Thigh: That part of the leg above the shank, and covered with feathers.

Thumb-marked comb: A single comb possessing indentations in the blade: a defect.

Ticked: Plumage tipped with a different colour, usually applied to V-shaped markings as in the Ancona. Also small coloured specks on any part of feathers of different colour from that of the ground colour.

Tipping: End of feathers tipped with a different coloured marking.

Top colour: see 'Surface colour'.

Top knot: see 'Crest'.

Glossary

Tri-coloured: Of three colours. The term refers chiefly to buff and red fowls, and is generally aplied only to males when their hackles and tails are dark compared with the general plumage, and the wing bows are darker; a fault.

Trio: A male and two females.

Tiple comb: see 'Pea comb'.

Twisted comb: A faulty-shaped pea or single comb.

Twisted feather: The shaft and web of the feather are twisted out of shape.

Type: Mould or shape (see 'Symmetry').

Undercolour: Colour seen when a bird is handled – that is, when the feathers are lifted; colour of fluff of feathers.

V-shaped comb: see 'Horn comb'.

Variety: A definite branch of a breed known by its distinctive colour or marking – for example, the black is a variety of the Leghorn. Sub-variety, a sub-division of an established variety, differing in shape of comb from the original – for example, the rose-combed black is a sub-variety of the black Leghorn. Thus the breed includes all the varieties and sub-varieties which would conform to the same standard type.

Vulture hocks: Stiff projecting quill feathers at the hock joint, growing on the thighs and extending backwards.

Walnut comb: see 'Strawberry comb'.

Wattles: The fleshy appendages at each side of base of beak, more strongly developed in male birds.

Web: A flat and thin structure. Web of feather: the flat or plume portion. Web of feet: the flat skin between the toes. Web of wing: the triangular skin seen when the wing is extended.

Whiskers: Feathers growing from the sides of the face (see 'Beard' and 'Muff').

Wing bar: Any line of dark colour across the middle of the wing, caused by the colour or marking of the feathers known as the lower wing coverts.

Wing bay: The triangular part of the folded wing between the wing bar and the point (see 'Diamond').

Wing bow: The upper or shoulder part of the wing.

Glossary

Wing butt or Wing point: The end of the primaries; the corners or ends of the wing. The upper ends are more properly called the shoulder butts and are thus termed by Game fanciers. The lower, similarly, are often called the lower butts.

Wing coverts: The feathers covering the roots of the secondary quills.

Work: The small spikes or working on top of a rose comb.

Wry back: A distorted bone structure usually causing a hump-backed condition.

Wry tail: A tail carried awry, to the right or left side of the continuation of the backbone, and not straight with the body of the fowl.

Sitters and non-sitters

Generally speaking these divide themselves – *heavy* breeds being sitters and the *light* breeds non-sitters – the former comprising mainly American and Asiatic breeds with Indian Game, Sussex, and Dorkings of British origin, while the latter are generally of Mediterranean origin. Unlike certain other countries, however, which classify three categories – *light, medium* and *heavy* – Great Britain adheres to two classes only. There are, therefore, certain exceptions to the foregoing generalization. The breeds are divided as follows:

Sitters

Asil, Australorp, Barnevelder, Brahma, Cochin, Crève-Cœur, Croad Langshan, Dorking, Faverolles, Frizzle, Houdan, Indian Game, Ixworth, Jersey Giant, Jubilee Indian Game, La Flèche, Malay, Marans, Marsh Daisy, Modern Game, Modern Langshan, New Hampshire Red, Norfolk Grey, North Holland Blue, Old English Game, Orloff, Orpington, Plymouth Rock, Rhode Island Red, Scots Dumpy, Silkie, Sultan, Sumatra Game, Sussex, Wyandotte, Yokohama.

Non-sitters

Ancona, Andalusian, Bresse, Campine, Hamburgh, Lakenvelder, Leghorn, Minorca, Old English Pheasant Fowl, Poland, Redcap, Scots Grey, Sicilian Buttercup, Spanish, Welsummer.

Bantams take the same classification, i.e. heavy or light, as their large prototypes.

In ducks the Indian Runner is probably the only real light breed, Campbells and Orpingtons being regarded as mediums and Aylesburies and Pekins as heavies.

There are no divisions in geese and turkeys. The trend in the latter is, however, to develop smaller carcases with broader breasts, hence the growing demand for breeds like Small Whites which give 3.6–4.6 kg (8–10 lb) birds at reasonably early ages.

Defects and deformities

The Poultry Club instructs its judges to work continuously for stock improvement in making their awards. They should keep in mind, therefore, the suitability of exhibits for the breeding pen, penalizing those defects that affect reproductive values or detract from what may be regarded as the highest merits of such birds.

Uniformity in judging and exhibiting is sought, the aim being to make show-pens reliable 'shop windows', displaying birds that buyers may claim with every confidence in their soundness and reproductive ability.

To be passed or penalized

The following are given as deformities and defects for which judges must pass or penalize an exhibit according to the seriousness of the defect.

Head points

Crossed or deformed beak. Malformation of beak. Badly dished bill in ducks. Blindness. Defective eyesight. Defective pupils. Odd eye colour. Comb that closes the nostrils. Side sprigs or double end on a single comb. Excessive fall of rosecomb on a single base. Split combs at blade. Fall-over

Crossed beak   Open beak   Sunken eye

Defects and Deformities

comb that obstructs vision. Malformed combs. Defective serrations. Peculiar head carriage. White in face. Wry neck. Indications of brain or nerve affection. Badly distended and sagging crop.

Dished bill

Back

Any deformity. Rounded or curved spine. Weak back formation. One bone higher than the other giving the back a lopsided appearance.

Bone structure

Pigeon breast. Seriously deformed breast bone. Malformation of breast bone that interferes with the internal organs. Down behind and curved end of breast bone which leads to drooping abdomen. Dented breast bone from perching. Enlargement on breast bone of turkey. Broken or malformed pelvic bones. Faulty stance.

Wings

Badly twisted or curled wing feathers. Slipped or drooping wing. Split wing, in a serious form, with large gap between primaries and secondaries. Defective wing formation in waterfowl. Slightly defective wing formation, even if well positioned and carried, to be penalized.

Roach back

Cut-away breast

Defects and Deformities

Left Ingrown leader.

Right Short of leader, and uneven wattles.

Left Rose comb falling to side and blocking vision.

Right Bad leader and coarse worked comb.

Left Beefy, and with part of blade too far forward.

Right Badly curved at rear end with spikes falling over.

Left Thumb mark and side sprig.

Right Flyaway comb.

Left Double folded comb.

Right Flop comb blocking vision.

Comb faults

Defects and Deformities

Tail

Wry. Squirrel. Defective parson's nose. Split or divided tail feathers or badly twisted feathers in tail. High tail in excess.

Legs and feet

Enlarged bone. Curved thigh bones. Malformation of bone. Bow legs or 'out of hocks'. Badly in at hocks. Duck toes. Crooked toes. Turned toes. Twisted feet. Enlarged toe joints. Lack of spurs on adult male. Leg feathers on clean legged birds.

Feathering

Soft or frizzled feathering in plain feathered breeds. Curled feathers on any part of body, including neck. Signs of slow feathering.

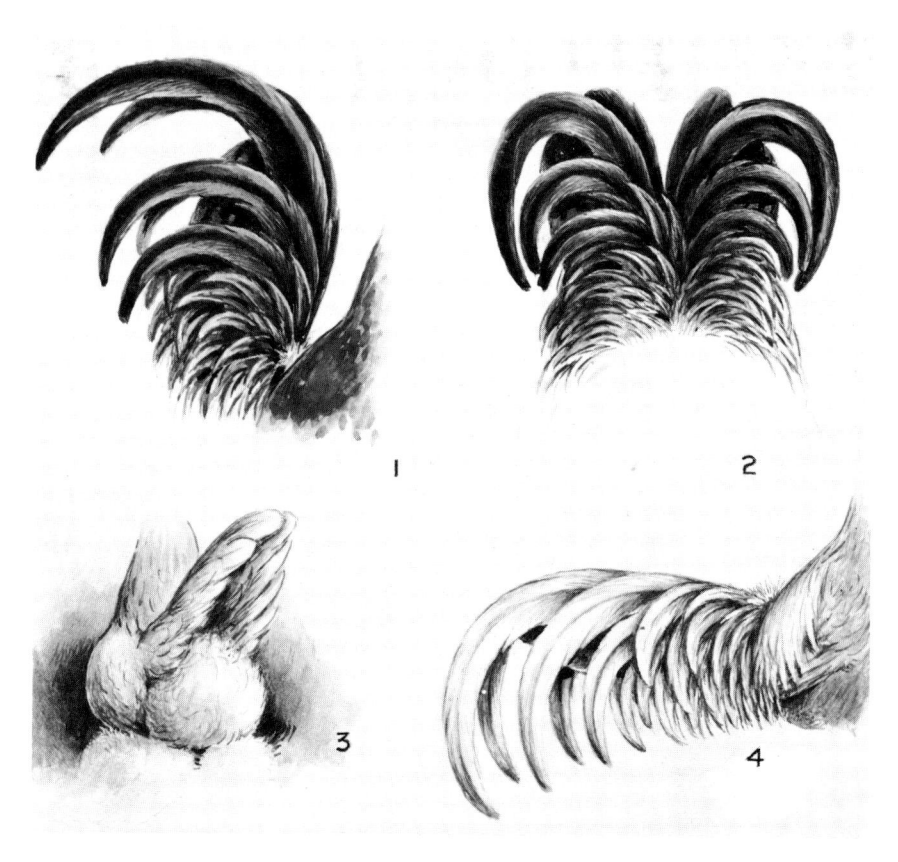

1. Squirrel tail   2. Split tail   3. Wry tail   4. Dropped tail

Defects and Deformities

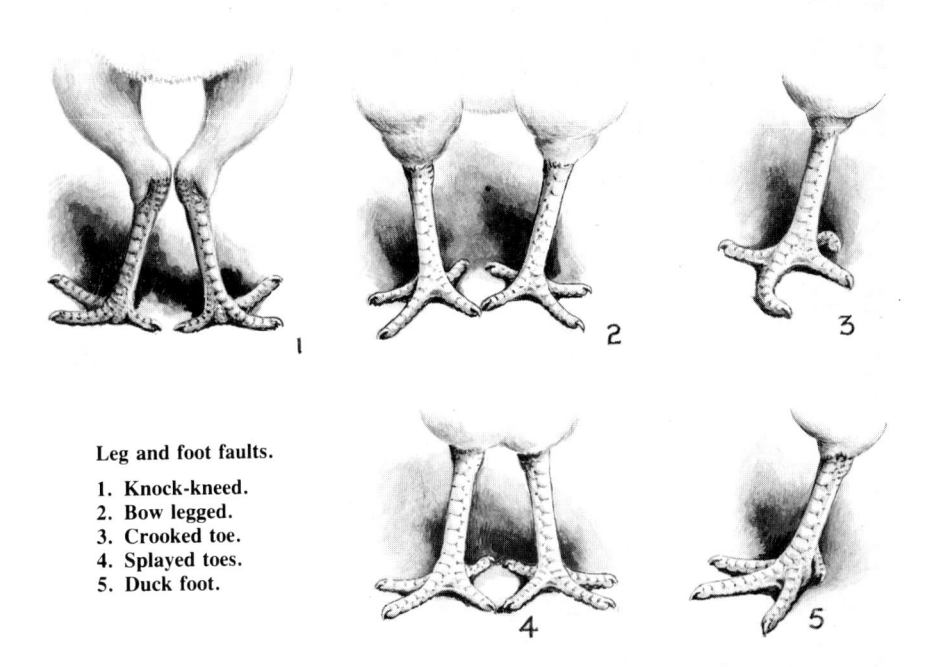

Leg and foot faults.

1. Knock-kneed.
2. Bow legged.
3. Crooked toe.
4. Splayed toes.
5. Duck foot.

Disease

Any disease or disorder not making for maximum health, condition, vigour and breeding fitness, including colds, heartiness, abdominal dropsy, cysts, egg substance in oviduct or abdomen. Sour crop. Impacted crop. Any other disease symptom or deformity.

Lack of breed characteristics

Any exhibit deficient in breed characteristics, so that it is an unworthy specimen of the breed or variety intended, must be passed.

Disqualifications

Any bird that in the opinion of the judge has been faked or tampered with shall be disqualified.